SOLVING AMERICA'S PROBLEMS: A CALL FOR UNITY AND ACTION

I0416154

PART 2:
Healthcare Reform:
Solutions for Improving
Access and Affordability

BY
HENRY E. PARKINS

1

COPYRIGHT PAGE

TABLE OF CONTENTS

INTRODUCTION

In the United States, healthcare reform stands as a perennial challenge, a labyrinthine maze of policies, stakeholders, and deeply entrenched interests. Access to quality healthcare and its affordability remain elusive aspirations for millions of Americans, casting shadows of uncertainty and vulnerability over their lives. From rural communities struggling to access basic services to urban populations burdened by exorbitant costs, the quest for a more equitable and efficient healthcare system resonates deeply across the nation.

This book, "Healthcare Reform: Solutions for Improving Access and Affordability in the US," embarks on a journey to dissect the complexities of America's healthcare landscape and to illuminate pathways toward meaningful change. It is an exploration of the multifaceted issues that plague the current system and a rallying cry for innovation, collaboration, and bold policy reform.

As we delve into the heart of this discourse, it becomes abundantly clear that the challenges facing healthcare in the US are not merely academic or

7

theoretical. They are lived experiences etched into the narratives of families grappling with impossible choices, of individuals confronting daunting barriers to care, and of communities striving for dignity and well-being.

In this introductory chapter, we set the stage for our exploration, laying bare the contours of the healthcare crisis and outlining the urgent imperative for reform. We survey the historical backdrop against which the current system has evolved, tracing the arc of progress and setbacks that have shaped its contours. We confront the stark realities of access disparities, affordability crises, and the profound human toll exacted by a system rife with inefficiencies and inequities.

Yet, amidst the challenges, there exists a glimmer of hope a recognition that the status quo is untenable, that change is not only possible but imperative. Across the spectrum of healthcare stakeholders patients, providers, policymakers, advocates, and innovators a groundswell of momentum is building, fueled by a collective determination to forge a better future for all Americans.

In the chapters that follow, we embark on a comprehensive exploration of the issues at hand, drawing upon insights from research, practice, and lived experience. We examine the structural barriers that hinder access to care, the economic forces that drive spiraling costs, and the policy interventions that hold promise for transformation. We spotlight innovative models of care delivery, the transformative potential of technology, and the imperative of addressing social determinants of health.

But our journey does not end with diagnosis and analysis. It extends to the realm of action to the corridors of power, the halls of governance, and the communities on the front lines of change. It is a call to arms for policymakers to transcend partisan divides and craft bold, visionary solutions. It is a call for healthcare providers to embrace the principles of patient-centered care and equitable access. And it is a call for all citizens to engage, to advocate, and to demand a healthcare system that reflects our highest ideals of compassion, justice, and solidarity.

As we embark on this odyssey of exploration and discovery, let us heed the voices of the marginalized, the vulnerable, and the forgotten. Let us honor the imperative of justice, the imperative of dignity, and the imperative of healing. For in the crucible of our shared humanity, the promise of healthcare reform beckons a promise of a future where health is not a privilege to be purchased but a right to be cherished, where access is not a barrier to be overcome but a bridge to be crossed, and where affordability is not a distant dream but a lived reality for all.

Definition and Importance of Healthcare Reform

Healthcare reform encompasses a broad spectrum of initiatives, policies, and interventions aimed at improving the accessibility, quality, and affordability of healthcare services within a given healthcare system. It represents a concerted effort to address systemic deficiencies, rectify inequities, and enhance the overall well-being of individuals and communities.

At its core, healthcare reform seeks to address the myriad challenges that plague healthcare systems, including barriers to access, escalating costs, disparities in care delivery, and inefficiencies in resource allocation. It encompasses a diverse array of strategies, ranging from legislative and regulatory changes to innovative care delivery models and technological advancements.

The importance of healthcare reform cannot be overstated, particularly in the context of the United States, where the healthcare system faces a myriad of complex and interconnected challenges. Access to affordable, high-quality healthcare is not merely a matter of individual well-being; it is a fundamental human right and a cornerstone of a just and equitable society.

Without meaningful reform, millions of Americans continue to face barriers to accessing essential healthcare services, perpetuating cycles of illness, poverty, and despair. The financial burden of healthcare costs threatens the economic security of families and undermines the nation's fiscal health. Disparities in healthcare access and outcomes exacerbate existing social inequities, perpetuating cycles of disadvantage and marginalization.

Moreover, the status quo is unsustainable, both morally and economically. The

11

spiraling costs of healthcare threaten to consume an ever-growing share of national resources, crowding out investments in education, infrastructure, and other vital public goods. The fragmentation and inefficiencies inherent in the current system impede the delivery of coordinated, patient-centered care, exacerbating disparities and compromising health outcomes.

Against this backdrop, healthcare reform assumes paramount importance as a moral imperative and a pragmatic necessity. It represents an opportunity to realign incentives, promote innovation, and build a more resilient, equitable healthcare system that serves the needs of all Americans.

Moreover, healthcare reform holds the potential to catalyze broader societal transformations, fostering a culture of health and well-being that transcends the confines of the healthcare system itself. By addressing social determinants of health, promoting preventive care, and fostering community partnerships, healthcare reform can lay the groundwork for healthier, more prosperous communities and a more vibrant, inclusive society.

In summary, healthcare reform is not merely a policy objective; it is a moral imperative, a societal imperative, and a pathway to a brighter, more equitable future for all Americans. By harnessing the power of innovation, collaboration, and collective action, we can chart a course toward a healthcare system that embodies our highest ideals of compassion, justice, and solidarity.

Historical Context of Healthcare Reform in the US

The quest for healthcare reform in the United States is deeply rooted in the nation's history, marked by a complex interplay of political, social, economic, and ideological forces. From the earliest days of the republic to the present day, the struggle to ensure access to affordable, high-quality healthcare has been a defining feature of American society.

The origins of healthcare reform in the US can be traced back to the late 19th and early 20th centuries, a period characterized by rapid industrialization, urbanization, and social upheaval. As millions of Americans migrated from rural areas to cities in search of economic opportunity, they were confronted with a healthcare system ill-equipped to meet their needs.

During this time, healthcare was largely decentralized and unregulated, with medical care delivered primarily through private practitioners, charitable organizations, and voluntary associations. Access to healthcare was often contingent on one's ability to pay, leaving millions of Americans without access to essential services.

The Great Depression of the 1930s laid bare the shortcomings of the existing healthcare system, as millions of Americans found themselves without access to medical care in the midst of economic crisis. It was during this tumultuous period that calls for healthcare reform began to gain traction, culminating in the passage of landmark legislation such as the Social Security Act of 1935, which laid the foundation for the modern welfare state and provided support for public health initiatives.

In the decades that followed, efforts to expand access to healthcare continued to evolve, driven by a growing recognition of healthcare as a fundamental human right.

14

The post-World War II era witnessed the emergence of employer-sponsored health insurance as a primary mechanism for accessing medical care, a trend that was further reinforced by the passage of the Internal Revenue Code of 1954, which provided tax incentives for employer-provided health benefits.

However, the reliance on employer-sponsored insurance left millions of Americans without coverage, particularly those who were unemployed or employed in low-wage industries. Calls for comprehensive healthcare reform grew louder throughout the latter half of the 20th century, fueled by rising healthcare costs, widening disparities in access, and growing public awareness of the inadequacies of the existing system.

The culmination of these efforts came with the passage of the Patient Protection and Affordable Care Act (ACA) in 2010, a landmark piece of legislation aimed at expanding access to healthcare coverage, controlling healthcare costs, and improving the quality of care. The ACA represented the most significant overhaul of the US healthcare system in decades, ushering in a new era of healthcare reform characterized by expanded insurance coverage, consumer protections, and investments in preventive care and public health.

Despite the progress achieved through the ACA, challenges remain in ensuring access to affordable, high-quality healthcare for all Americans. Rising healthcare costs, persistent disparities in access and outcomes, and political polarization continue to pose formidable obstacles to comprehensive healthcare reform.

As we confront the complexities of the present moment, it is imperative that we draw upon the lessons of history and the spirit of collective action that has propelled previous generations toward progress. By embracing innovation, collaboration, and a commitment to equity, we can build upon the foundation laid by past reform efforts and forge a brighter, more inclusive future for healthcare in America.

Overview of Current Challenges in Access and Affordability

In the landscape of healthcare in the United States, access to quality care and its affordability stand as paramount challenges, presenting formidable barriers to the well-being and security of millions of Americans. As we navigate the complexities of the present moment, several pressing challenges underscore the urgent need for comprehensive healthcare reform.

Access Disparities: Despite advances in medical science and technology, significant disparities persist in access to healthcare services across demographic groups and geographic regions. Rural communities, in particular, face acute shortages of healthcare providers and facilities, limiting residents' ability to access timely and appropriate care. Additionally, marginalized populations, including racial and ethnic minorities, individuals with disabilities, and low-income individuals, often encounter

17

systemic barriers that impede their access to essential services.

Insurance Coverage Gaps:

While the Affordable Care Act (ACA) expanded access to health insurance coverage for millions of Americans, significant coverage gaps remain. Many individuals and families remain uninsured or underinsured, lacking access to essential health benefits and preventive services. Furthermore, the proliferation of high-deductible health plans and rising out-of-pocket costs place a heavy financial burden on individuals and families, deterring them from seeking necessary care and exacerbating health disparities.

Escalating Healthcare Costs:

Healthcare costs in the United States continue to escalate at an unsustainable pace, outpacing inflation and consuming an ever-growing share of national resources. Factors contributing to rising healthcare costs include the high prices of medical services and prescription drugs, administrative inefficiencies, and the proliferation of costly technologies and treatments. The financial burden of healthcare costs not only strains

18

household budgets but also imposes significant fiscal pressures on businesses, government programs, and the broader economy.

Fragmentation and Inefficiencies in Care Delivery:

The fragmentation of the healthcare system, characterized by siloed care delivery, fragmented payment models, and disjointed care coordination, undermines the quality and efficiency of healthcare delivery. Patients often navigate complex and disjointed care pathways, leading to gaps in care, medical errors, and suboptimal health outcomes. Additionally, administrative burdens and bureaucratic inefficiencies divert valuable resources away from patient care, contributing to the overall inefficiency of the healthcare system.

Social Determinants of Health:

Recognizing the profound influence of social and environmental factors on health outcomes, healthcare reform efforts must address the social determinants of health, including poverty, housing instability, food insecurity, and lack of access to education and employment opportunities. These upstream determinants significantly impact individuals' health status and healthcare utilization patterns, necessitating a holistic and integrated approach to healthcare delivery and resource allocation.

In confronting these challenges, it is imperative that healthcare reform efforts prioritize equity, accessibility, and affordability, centering the needs and experiences of individuals and communities most adversely affected by systemic inequities. By fostering innovation, collaboration, and a commitment to social justice, we can chart a course toward a more equitable and inclusive healthcare system—one that ensures access to high-quality care for all Americans, regardless of their socioeconomic status, race, ethnicity, or geographic location.

CHAPTER 1

UNDERSTANDING THE CURRENT HEALTHCARE LANDSCAPE

In navigating the complexities of healthcare reform in the United States, it is essential to grasp the multifaceted nature of the current healthcare landscape. Shaped by historical precedent, socio-economic dynamics, and evolving healthcare paradigms, the contemporary healthcare system reflects a mosaic of stakeholders, institutions, and policy frameworks.

Diverse Stakeholders: At the heart of the healthcare landscape are a diverse array of stakeholders, each with distinct roles, interests, and perspectives. These stakeholders include patients and families, healthcare providers (such as physicians, nurses, and allied health professionals), insurers, employers, government agencies, pharmaceutical companies, advocacy organizations, and community-based health entities. Understanding the motivations, incentives, and power dynamics that shape the interactions among these stakeholders is critical to navigating the complexities of healthcare reform.

Fragmented Delivery Systems:
The delivery of healthcare services in the United States is characterized by fragmentation and heterogeneity, with a multitude of providers, settings, and modalities involved in care delivery. From hospitals and primary care clinics to specialty practices and community health centers, patients navigate a complex maze of care options, each with its own strengths, limitations, and costs. Achieving seamless care coordination and integration across disparate care settings remains a significant challenge, often resulting in gaps in care, duplication of services, and suboptimal health outcomes.

Payment and Financing Mechanisms:
The financing of healthcare in the United States is marked by a patchwork of payment mechanisms and reimbursement models, each with its own set of incentives and disincentives. Traditional fee-for-service arrangements, which reimburse providers based on the

volume of services delivered, have long been criticized for incentivizing unnecessary care and driving up costs. In recent years, value-based payment models and alternative payment arrangements have gained traction, incentivizing providers to deliver high-quality, cost-effective care and prioritize preventive services and population health management.

Regulatory and Policy Frameworks:

The regulation of healthcare in the United States is governed by a complex web of federal, state, and local laws, regulations, and policies. Key regulatory bodies include the Centers for Medicare & Medicaid Services (CMS), the Food and Drug Administration (FDA), and state-level departments of health. The passage of landmark legislation such as the Affordable Care Act (ACA) and the Health Insurance Portability and Accountability Act (HIPAA) has reshaped the healthcare landscape, expanding access to coverage, enhancing consumer protections, and promoting value-based care delivery.

23

Technological Advancements:

Rapid advancements in technology such as electronic health records (EHRs), telemedicine, precision medicine, and artificial intelligence have transformed the delivery and management of healthcare services. These technologies hold the promise of improving care coordination, enhancing diagnostic accuracy, and empowering patients to actively participate in their own care. However, challenges related to data privacy, interoperability, and equity in access to technology persist, necessitating thoughtful policy and regulatory responses.

By comprehensively understanding the intricate interplay of these factors, stakeholders can identify opportunities for innovation, collaboration, and systemic change. Effective healthcare reform requires a nuanced understanding of the current healthcare landscape a landscape shaped by historical precedent, shaped by historical precedent, socio-economic dynamics, and evolving healthcare paradigms. Through informed dialogue, strategic partnerships, and a shared commitment to equity and justice, we can forge a more equitable, accessible, and sustainable healthcare system that meets the needs of all Americans.

Analysis of the US Healthcare System

The United States healthcare system is a complex ecosystem shaped by a multitude of factors, including historical precedent, socio-economic dynamics, political influences, and technological advancements. While the US healthcare system is characterized by innovation, cutting-edge medical technologies, and world-renowned institutions, it also faces profound challenges that undermine its ability to provide accessible, affordable, and equitable care to all Americans.

Fragmentation and Complexity:

One of the defining features of the US healthcare system is its fragmentation and complexity. Patients navigate a fragmented landscape of healthcare providers, insurers, and care settings, each with its own protocols, payment mechanisms, and administrative requirements. This fragmentation contributes to inefficiencies, administrative burdens, and barriers to care coordination, ultimately undermining the quality and continuity of care.

25

High Costs and Spending:

Healthcare costs in the United States are among the highest in the world, far outpacing those of other developed nations. The US spends a disproportionately large share of its GDP on healthcare, yet lags behind other countries in key indicators of health outcomes, such as life expectancy and infant mortality. The drivers of high healthcare costs include administrative overhead, the high prices of medical services and prescription drugs, defensive medicine practices, and a fee-for-service payment model that incentivizes volume over value.

Access Disparities:

Despite advances in medical science and technology, significant disparities persist in access to healthcare services across demographic groups and geographic regions. Millions of Americans remain uninsured or underinsured, lacking access to essential health benefits and preventive services. Vulnerable populations, including racial and ethnic minorities, individuals with disabilities, and low-income individuals, face disproportionate barriers to accessing care, exacerbating health

26

disparities and perpetuating cycles of disadvantage.

Inequities in Health Outcomes:

The United States grapples with persistent inequities in health outcomes, with marginalized communities experiencing higher rates of chronic diseases, premature mortality, and preventable hospitalizations. Social determinants of health, including poverty, inadequate housing, food insecurity, and lack of access to education and employment opportunities, significantly impact individuals' health status and healthcare utilization patterns. Addressing these upstream determinants is essential to achieving health equity and improving overall population health.

Lack of Emphasis on Prevention and Primary Care: The US healthcare

system has historically placed greater emphasis on acute and specialty care services than on preventive care and primary care. This imbalance contributes to the high prevalence of preventable chronic diseases and the overutilization of costly healthcare services. Investing in preventive care, health promotion, and primary care infrastructure is critical to reducing

27

healthcare costs, improving health outcomes, and fostering a more sustainable healthcare system.

In confronting these challenges, healthcare reform efforts must prioritize equity, accessibility, and affordability, centering the needs and experiences of individuals and communities most adversely affected by systemic inequities. This necessitates a comprehensive approach that addresses the root causes of health disparities, promotes value-based care delivery, enhances care coordination, and fosters innovation in healthcare delivery and payment models.

By harnessing the power of collaboration, innovation, and evidence-based policymaking, we can chart a course toward a more equitable, accessible, and sustainable healthcare system one that reflects our shared commitment to health equity, social justice, and the well-being of all Americans.

Key Stakeholders and Their Roles in Healthcare Reform:

Patients and Families:

Role: Patients and their families are at the center of the healthcare system, and their experiences, needs, and perspectives are paramount in shaping healthcare reform efforts. Patients advocate for improved access to care, affordability, and quality services. They also play a crucial role in shared decision-making and care coordination.

Healthcare Providers:

Role: Healthcare providers, including physicians, nurses, allied health professionals, and pharmacists, deliver direct patient care and are instrumental in implementing healthcare reform initiatives. Providers advocate for evidence-based practices, patient-centered care, and professional autonomy. They also contribute to quality improvement efforts and clinical innovation.

Healthcare Administrators and Executives:

Role: Healthcare administrators and executives oversee the management and operations of healthcare organizations, including hospitals, clinics, and health systems. They are responsible for strategic planning, resource allocation, and financial management. Administrators play a critical role in implementing healthcare reform policies, optimizing organizational performance, and fostering a culture of continuous improvement.

Insurers and Payers:

Role: Insurers and payers, including private insurance companies, government programs (such as Medicare and Medicaid), and employer-sponsored health plans, play a central role in financing healthcare services and managing risk. Insurers negotiate payment rates with providers, administer benefits, and develop coverage policies. They also advocate for policies that promote cost containment and quality improvement.

Government Agencies and Policymakers:

Role: Government agencies at the federal, state, and local levels play a pivotal role in shaping healthcare policy, regulation, and financing. Policymakers craft legislation, regulations, and reimbursement policies that govern the delivery and payment of healthcare services. Government agencies also oversee public health initiatives, healthcare quality improvement programs, and research funding.

Healthcare Advocacy Organizations:

Role: Healthcare advocacy organizations represent the interests of patients, consumers, providers, and other stakeholders in the healthcare system. These organizations advocate for policy reforms that promote access to care, affordability, and quality services. They engage in public education, grassroots mobilization, and lobbying efforts to advance their policy priorities.

Employers and Business Groups:

Role: Employers and business groups play a significant role in providing healthcare

31

coverage to employees and advocating for policies that address healthcare costs and quality. Employers negotiate benefit packages, contribute to insurance premiums, and promote wellness initiatives in the workplace. Business groups advocate for policies that support a competitive healthcare marketplace and promote economic growth.

Healthcare Technology and Innovation Companies:

Role: Healthcare technology and innovation companies develop and deploy technologies, digital solutions, and medical devices that enhance care delivery, improve patient outcomes, and streamline administrative processes. These companies drive innovation in areas such as telemedicine, electronic health records, data analytics, and medical devices. They collaborate with stakeholders to integrate technology into healthcare delivery and support evidence-based decision-making.

Community Organizations and Advocates:

Role: Community organizations and advocates work at the grassroots level to address social determinants of health,

32

promote health equity, and empower underserved populations. These organizations provide health education, outreach, and support services to vulnerable communities. They advocate for policies that address disparities in access to care, housing, education, and employment.

By engaging these key stakeholders in collaborative dialogue, partnership, and shared decision-making, healthcare reform efforts can effectively address the complex challenges of improving access, affordability, and quality in the US healthcare system.

Factors Contributing to Access and Affordability Issues

Lack of Health Insurance Coverage: One of the primary barriers to accessing healthcare services in the United States is the lack of health insurance coverage. Millions of Americans remain uninsured, often due to prohibitive costs, eligibility criteria, or lack of awareness about available coverage options. Without insurance, individuals may forgo preventive care, delay necessary treatments, and face financial hardship in the event of medical emergencies.

High Premiums, Deductibles, and Copayments: Even for individuals with health insurance coverage, high premiums, deductibles, and copayments can pose significant financial barriers to accessing care. Rising healthcare costs have led to increased out-of-pocket expenses for patients, making it difficult for many individuals and families to afford necessary medical services and medications.

Limited Access to Primary Care Providers: Access to primary care providers, including physicians, nurse practitioners, and physician assistants, remains limited for many Americans, particularly those in rural and underserved communities. Shortages of primary care providers, long wait times for appointments, and geographic disparities in healthcare infrastructure contribute to barriers in accessing timely and preventive care.

Geographic and Transportation Barriers:

Geographic barriers, such as living in rural or remote areas, can impede access to healthcare services due to limited availability of providers and facilities. Transportation challenges, including lack of public transportation options and long travel distances to reach healthcare facilities, further exacerbate access disparities for individuals living in rural and low-income communities.

Provider Shortages and Workforce Challenges:

Shortages of healthcare providers, particularly in primary care and mental health specialties, contribute to access issues in many parts of the country. Factors such as aging provider demographics, workforce maldistribution, and limited training opportunities for healthcare professionals further exacerbate provider shortages and constrain access to care for underserved populations.

Language and Cultural Barriers:

Language and cultural barriers can pose significant challenges to accessing healthcare services for individuals with limited English proficiency or from diverse cultural backgrounds. Lack of access to interpretation services, culturally competent care, and culturally sensitive healthcare settings can impede effective communication and diminish the quality of care for patients from linguistic and cultural minority groups.

Systemic Disparities and Discrimination: Systemic disparities in access to healthcare services persist across racial, ethnic, socioeconomic, and demographic groups. Structural racism, discrimination, and implicit bias within the healthcare system contribute to disparities in health outcomes, quality of care, and access to specialty services. Efforts to address systemic inequities and promote health equity are essential components of healthcare reform initiatives.

Market Forces and Profit Incentives: Market forces and profit incentives within the healthcare industry can drive up costs, prioritize revenue generation over patient care, and create barriers to affordable access. Pharmaceutical pricing practices, consolidation within the healthcare

market, and profit-driven decision-making by insurers and providers can contribute to affordability issues and limit access to essential treatments and services.

By addressing these underlying factors contributing to access and affordability issues, healthcare reform efforts can strive to create a more equitable, accessible, and affordable healthcare system that meets the needs of all Americans. This requires comprehensive policy solutions, investment in healthcare infrastructure, workforce development initiatives, and efforts to promote health equity and social justice.

CHAPTER 2

ACCESS TO HEALTHCARE

Access to healthcare is a fundamental human right and a cornerstone of public health and social justice. It encompasses the ability of individuals to obtain timely, appropriate, and affordable healthcare services when needed, without encountering barriers related to cost, geography, or social determinants of health. In the United States, ensuring equitable access to healthcare remains a critical challenge, with millions of Americans facing barriers that prevent them from obtaining necessary medical care and services.

Insurance Coverage Gaps: A significant barrier to access to healthcare in the United States is the presence of insurance coverage gaps. While the Affordable Care Act (ACA) expanded access to health insurance coverage for millions of Americans, a substantial number of individuals remain uninsured or underinsured. Those without insurance coverage often delay or forgo necessary medical care due to concerns about affordability and financial hardship.

Cost of Care: The high cost of healthcare services poses a significant barrier to access for many Americans. Rising healthcare costs, including premiums, deductibles, copayments, and out-of-pocket expenses, can create financial barriers that prevent individuals and families from seeking necessary medical care. The fear of incurring medical debt or facing bankruptcy due to healthcare expenses can deter individuals from accessing preventive care and timely treatments.

Geographic Barriers: Access to healthcare services is often limited by geographic barriers, particularly in rural and underserved areas. Shortages of healthcare providers, including primary care physicians, specialists, and mental health professionals, can result in long wait times for appointments and limited availability of services. In rural communities, the distance to healthcare facilities and lack of transportation options further exacerbate access challenges.

Provider Shortages and Workforce Maldistribution:

Shortages of healthcare providers, particularly in primary care and mental health specialties, contribute to access disparities in many parts of the country. Workforce maldistribution, wherein healthcare providers are concentrated in urban areas while rural and underserved communities face shortages, limits access to care for individuals living in remote or medically underserved regions.

Language and Cultural Barriers:

Language and cultural barriers can hinder access to healthcare services for individuals with limited English proficiency or from diverse cultural backgrounds. Lack of access to interpretation services, culturally competent care, and culturally sensitive healthcare settings can impede effective communication and diminish the quality of care for patients from linguistic and cultural minority groups.

Systemic Disparities and Discrimination:

Systemic disparities in access to healthcare services persist across racial, ethnic, socioeconomic, and demographic groups. Structural racism, discrimination, and implicit bias within the healthcare system contribute to disparities in health outcomes, quality of care, and access to specialty services. Efforts to address systemic inequities and promote health equity are essential components of healthcare reform initiatives.

Addressing barriers to access to healthcare requires comprehensive policy solutions, investment in healthcare infrastructure, workforce development initiatives, and efforts to promote health equity and social justice. By prioritizing access to care for all Americans, regardless of their socioeconomic status, race, ethnicity, or geographic location, healthcare reform can advance the goal of creating a more equitable, accessible, and affordable healthcare system in the United States.

Barriers to Access for Different Demographics: Rural Populations

Geographic Distance: Rural populations often face significant barriers to accessing healthcare due to the geographic distance to healthcare facilities. Limited availability of healthcare providers and facilities in rural areas can result in long travel times and transportation challenges for individuals seeking medical care.

Provider Shortages: Shortages of healthcare providers, including primary care physicians, specialists, and mental health professionals, are common in rural communities. Limited access to healthcare professionals exacerbates access disparities and can result in long wait times for appointments and limited availability of services.

Healthcare Infrastructure: Rural areas may lack essential healthcare infrastructure, including hospitals, clinics, and diagnostic facilities. Limited access to emergency services, specialty care, and medical technology further hinders rural residents' ability to obtain timely and comprehensive healthcare services.

Low-Income Individuals:

Financial Barriers: Low-income individuals often face significant financial barriers to accessing healthcare services due to the high cost of care. Even with

insurance coverage, out-of-pocket expenses such as deductibles, copayments, and prescription drug costs can be prohibitive for individuals with limited financial resources.

Lack of Insurance Coverage:

Uninsured or underinsured individuals are particularly vulnerable to barriers to access due to the inability to afford health insurance premiums or lack of eligibility for public insurance programs such as Medicaid. Without insurance coverage, individuals may delay or forgo necessary medical care, leading to adverse health outcomes and increased healthcare costs in the long term.

Limited Access to Preventive

Care: Low-income individuals often have limited access to preventive care services such as screenings, vaccinations, and wellness exams. Without regular preventive care, individuals are more likely to experience untreated health conditions, exacerbating disparities in health outcomes and increasing the burden on the healthcare system.

Racial and Ethnic Minorities:

Language and Cultural Barriers:

Racial and ethnic minorities may encounter language and cultural barriers that impede access to healthcare services. Limited access to interpretation services, culturally competent care, and healthcare providers from diverse backgrounds can hinder effective communication and diminish the quality of care for minority patients.

Discrimination and Bias: Structural racism, discrimination, and implicit bias within the healthcare system can contribute to disparities in access to care and health outcomes for racial and ethnic minority populations. Minority individuals may experience differential treatment, barriers to care coordination, and unequal access to specialty services based on their race or ethnicity.

Socioeconomic Disparities: Racial

and ethnic minority populations are disproportionately affected by socioeconomic disparities, including poverty, unemployment, and lack of access to education and housing. These social determinants of health can significantly impact individuals' ability to access

44

healthcare services and contribute to disparities in health outcomes among minority populations.

Addressing barriers to access for different demographic groups requires targeted interventions, including increasing healthcare workforce diversity, expanding access to insurance coverage, investing in healthcare infrastructure in underserved areas, and implementing policies to address social determinants of health. By addressing the unique needs and challenges faced by diverse populations, healthcare reform efforts can promote equity, accessibility, and affordability in the US healthcare system.

Impact of Insurance Coverage Gaps

Insurance coverage gaps, wherein individuals lack adequate health insurance coverage or remain uninsured altogether, have profound implications for access to healthcare and overall health outcomes in the United States. These coverage gaps contribute to a myriad of challenges that hinder individuals' ability to obtain timely and appropriate medical care, exacerbating health disparities and financial burdens across the population.

Delayed or Foregone Care: One of the most significant impacts of insurance coverage gaps is the delay or avoidance of necessary medical care. Uninsured individuals and those with limited coverage may postpone seeking medical attention for acute illnesses, chronic conditions, or preventive services due to concerns about cost. Delayed care can result in the progression of health conditions, exacerbation of symptoms, and increased healthcare costs in the long term.

Financial Hardship: Individuals who lack adequate insurance coverage may face significant financial hardship when accessing healthcare services. Without insurance, medical expenses such as hospitalizations, surgeries, medications, and diagnostic tests can quickly accumulate, leading to medical debt, bankruptcy, and financial instability for individuals and families. The burden of healthcare costs can disproportionately affect low-income individuals and those with chronic health conditions,

exacerbating socioeconomic disparities and perpetuating cycles of poverty.

Limited Access to Preventive Care:

Insurance coverage gaps often result in limited access to preventive care services, including screenings, vaccinations, wellness exams, and preventive counseling. Without coverage for preventive services, individuals are less likely to receive timely screenings and interventions that can detect health conditions at an early stage and prevent the progression of diseases. The absence of preventive care contributes to higher rates of preventable illnesses, chronic conditions, and avoidable hospitalizations among uninsured and underinsured populations.

Health Disparities and Unmet Needs:

Insurance coverage gaps exacerbate disparities in access to healthcare and health outcomes across demographic groups. Racial and ethnic minorities, low-income individuals, rural populations, and other vulnerable communities are disproportionately affected by coverage gaps and face barriers to accessing essential healthcare

47

services. Disparities in insurance coverage contribute to disparities in health outcomes, with uninsured and underinsured individuals experiencing higher rates of morbidity, mortality, and unmet healthcare needs compared to those with adequate insurance coverage.

Strain on the Healthcare System:

Insurance coverage gaps place strain on the healthcare system as uninsured and underinsured individuals seek care through emergency departments, safety-net clinics, and other acute care settings. The reliance on emergency care for non-emergent conditions contributes to overcrowding, longer wait times, and higher healthcare costs for hospitals and providers. The burden of uncompensated care, wherein healthcare providers absorb the costs of treating uninsured patients, further strains healthcare resources and compromises the financial viability of safety-net institutions.

Addressing insurance coverage gaps is a critical component of healthcare reform efforts aimed at improving access and affordability in the United States. Comprehensive strategies to expand access to affordable insurance coverage, increase enrollment in public programs such as Medicaid and the Children's Health Insurance Program (CHIP), and enhance consumer protections can help mitigate the impact of coverage gaps and promote equitable access to healthcare for all Americans. Additionally, efforts to address systemic inequities, social determinants of health, and barriers to care coordination are essential to achieving meaningful improvements in access and affordability across the healthcare system.

Case Study 1: Rural Healthcare Access Disparities

In rural Appalachia, access to healthcare is a significant challenge due to geographic isolation, provider shortages, and limited healthcare infrastructure. The case of Mrs. Johnson, a 65-year-old woman living in a remote mountain community, illustrates the barriers faced by rural populations in accessing essential healthcare services. Mrs. Johnson suffers from diabetes and requires regular monitoring of her blood sugar levels and access to prescription medications to manage her condition.

Despite her healthcare needs, Mrs. Johnson struggles to access adequate care due to the lack of healthcare providers in her area. The nearest primary care physician is located over 50 miles away, and Mrs. Johnson lacks reliable transportation to travel to appointments. Additionally, the local clinic has limited hours of operation and often faces shortages of medical supplies and medications.

As a result, Mrs. Johnson's diabetes goes unmanaged, leading to complications such as uncontrolled blood sugar levels, increased risk of cardiovascular disease, and decreased quality of life. Without access to preventive care and timely interventions, Mrs. Johnson's health deteriorates, exacerbating the burden on the healthcare system and compromising her overall well-being.

Case Study 2: Urban Healthcare Access Disparities

In inner-city neighborhoods, access to healthcare is hampered by socioeconomic disparities, language barriers, and inadequate healthcare infrastructure. The case of Mr. Garcia, a 40-year-old immigrant living in a low-income neighborhood, highlights the challenges faced by urban populations in accessing comprehensive healthcare services. Mr. Garcia suffers from hypertension and requires regular check-ups, medication management, and access to specialty care to monitor his condition.

However, Mr. Garcia encounters numerous barriers to accessing care in his community. Limited availability of primary care providers and long wait times for appointments at local clinics make it difficult for him to receive timely medical attention. Additionally, language barriers and cultural differences make it challenging for Mr. Garcia to communicate with healthcare providers and navigate the healthcare system effectively.

As a result, Mr. Garcia's hypertension remains poorly controlled, increasing his risk of complications such as heart attack, stroke, and kidney disease. Without access to culturally competent care and support services, Mr. Garcia's health outcomes suffer, perpetuating disparities in access and quality of care among underserved urban populations.

Case Study 3: Healthcare Access Disparities among Minority Communities

In communities of color, access to healthcare is influenced by systemic racism, discrimination, and disparities in socioeconomic status. The case of Mrs. Lee, a 30-year-old African American woman living in a low-income neighborhood, illustrates the intersectional challenges faced by minority populations in accessing healthcare services. Mrs. Lee suffers from depression and anxiety and requires access to mental health counseling, medication management, and support services to address her mental health needs.

However, Mrs. Lee encounters numerous barriers to accessing care in her community. Limited availability of mental health providers, stigma surrounding mental illness, and lack of insurance coverage for mental health services make it difficult for her to seek treatment. Additionally, systemic racism and discrimination within the healthcare system contribute to mistrust and reluctance among minority individuals to seek care.

As a result, Mrs. Lee's mental health deteriorates, leading to worsening symptoms, decreased functioning, and decreased quality of life. Without access to culturally competent care and support services, Mrs. Lee's mental health needs remain unmet, perpetuating disparities in access and outcomes among minority communities.

These case studies underscore the urgent need for healthcare reform efforts to address access disparities and promote equitable access to care for all Americans, regardless of their geographic location, socioeconomic status, or demographic background. Comprehensive strategies to expand access to healthcare services,

53

increase healthcare workforce diversity, and address social determinants of health are essential to achieving meaningful improvements in access and affordability across the healthcare system.

CHAPTER 3

AFFORDABILITY IN HEALTHCARE

Affordability is a critical dimension of healthcare access, encompassing the financial resources required for individuals and families to obtain necessary medical care without facing undue financial hardship. In the United States, concerns about healthcare affordability persist, as rising healthcare costs, out-of-pocket expenses, and insurance premiums strain household budgets and contribute to disparities in access to care. Understanding the challenges and drivers of healthcare affordability is essential for designing effective healthcare reform solutions aimed at improving access and reducing financial barriers for all Americans.

Rising Healthcare Costs: One of the primary drivers of healthcare affordability challenges is the persistent rise in healthcare costs. Healthcare expenditures in the United States continue to outpace inflation, driven by factors such

as technological advancements, prescription drug prices, administrative expenses, and increased utilization of medical services. As healthcare costs escalate, individuals and families face higher insurance premiums, deductibles, copayments, and out-of-pocket expenses, making it increasingly difficult to afford necessary medical care.

High Insurance Premiums and Deductibles: For individuals with health insurance coverage, high premiums and deductibles pose significant affordability challenges. Many Americans struggle to afford monthly insurance premiums, especially for employer-sponsored plans or marketplace insurance policies. High deductibles require individuals to pay a substantial amount out-of-pocket before insurance coverage kicks in, leading some individuals to delay or forgo necessary medical care due to concerns about cost.

Out-of-Pocket Expenses: Out-of-pocket expenses, including copayments, coinsurance, and prescription drug costs, contribute to the financial burden of healthcare for individuals and families. Prescription medications, in particular, can

be prohibitively expensive, especially for individuals with chronic conditions or rare diseases that require ongoing treatment. High out-of-pocket costs can deter individuals from seeking necessary medical care, leading to adverse health outcomes and increased healthcare costs in the long term.

Underinsured Population: Despite having health insurance coverage, many Americans remain underinsured, meaning their insurance plans do not adequately protect them from high out-of-pocket expenses or provide comprehensive coverage for necessary medical services. Underinsured individuals may face substantial financial burdens when accessing healthcare, leading to medical debt, bankruptcy, and financial insecurity.

Lack of Price Transparency: The lack of price transparency in the healthcare system makes it difficult for consumers to compare prices, shop for lower-cost providers, and make informed decisions about their healthcare spending. Without access to transparent pricing information, individuals may be unaware of the true cost of medical services and may

be surprised by unexpected medical bills, further exacerbating affordability challenges.

Social Determinants of Health:

Social determinants of health, including income, education, employment, and housing, significantly influence individuals' ability to afford healthcare services. Low-income individuals and those facing financial insecurity are more likely to experience affordability barriers and may delay or forgo necessary medical care due to concerns about cost. Addressing social determinants of health is essential for improving healthcare affordability and promoting health equity across diverse populations.

Addressing affordability challenges in healthcare requires comprehensive reform efforts aimed at reducing healthcare costs, increasing price transparency, expanding insurance coverage, and addressing social determinants of health. Policy solutions such as expanding access to affordable insurance options, implementing price transparency measures, controlling prescription drug prices, and investing in preventive care and public health initiatives can help mitigate affordability barriers and ensure that all Americans have access to high-quality, affordable healthcare services. By prioritizing affordability as a key component of healthcare reform, policymakers can advance the goal of creating a more equitable, accessible, and sustainable healthcare system for all.

Rising Healthcare Costs and Their Implications

Financial Burden on Individuals and Families: The primary implication of rising healthcare costs is the significant financial burden placed on individuals and families. As healthcare expenditures continue to outpace inflation, Americans face higher insurance premiums, deductibles, copayments, and out-of-pocket expenses for medical care. The increasing cost of healthcare can strain household budgets, leading to financial hardship, medical debt, and bankruptcy for individuals and families who struggle to afford necessary medical services.

Barriers to Access and Affordability: Rising healthcare costs create barriers to access and affordability for many Americans, particularly those who are uninsured, underinsured, or living on limited incomes. High insurance premiums and out-of-pocket expenses can deter individuals from seeking necessary medical care, leading to delays in treatment, worsening health outcomes, and increased healthcare costs in the long term. Affordability barriers disproportionately affect vulnerable populations, including low-income individuals, racial and ethnic minorities, and those with chronic health conditions.

Impact on Healthcare Utilization and Health Outcomes:
The escalating cost of healthcare can influence patterns of healthcare utilization and health outcomes. Individuals may forgo preventive care, routine screenings, and follow-up appointments due to concerns about cost, leading to missed opportunities for early detection and intervention. Delayed or deferred medical care can result in the progression of health conditions, increased morbidity, and avoidable hospitalizations, placing additional strain on the healthcare system

and compromising patients' health outcomes.

Financial Sustainability of the Healthcare System:

Rising healthcare costs pose challenges to the financial sustainability of the healthcare system as a whole. Healthcare expenditures account for a significant portion of the national economy, placing strain on government budgets, employer-sponsored health plans, and individual consumers. The increasing cost of healthcare contributes to fiscal pressures on public insurance programs such as Medicare and Medicaid, leading to debates about program solvency, reimbursement rates, and healthcare financing mechanisms.

Impact on Healthcare Providers and Institutions:

Healthcare providers and institutions also feel the impact of rising healthcare costs. Providers face pressure to contain costs, improve efficiency, and maximize revenue streams while maintaining high-quality patient care. Hospitals, clinics, and healthcare systems grapple with financial challenges, including uncompensated care, declining

61

reimbursement rates, and rising operating expenses. The financial strain on healthcare providers and institutions can affect staffing levels, resource allocation, and the availability of essential healthcare services in communities.

Need for Healthcare Reform: The implications of rising healthcare costs underscore the urgent need for comprehensive healthcare reform efforts aimed at addressing cost drivers, improving efficiency, and promoting value-based care delivery. Policy solutions such as expanding access to affordable insurance coverage, controlling prescription drug prices, investing in preventive care and population health management, and promoting transparency and accountability in healthcare pricing and reimbursement can help mitigate affordability barriers and ensure that all Americans have access to high-quality, affordable healthcare services.

By addressing the root causes of rising healthcare costs and implementing evidence-based reforms, policymakers, healthcare stakeholders, and community leaders can work together to create a more equitable, accessible, and sustainable healthcare system that meets the needs of individuals and families across the United States.

The Role of Insurance Premiums, Deductibles, and Copayments

Insurance Premiums:

Insurance premiums represent the amount individuals and families pay to maintain health insurance coverage. Premiums are typically paid on a monthly basis, either by individuals directly or through employer-sponsored health plans. The role of insurance premiums is essential in providing financial protection and access to healthcare services for insured individuals.

Affordability of insurance premiums is a critical factor in determining individuals' ability to obtain and maintain health insurance coverage. High premiums may pose affordability barriers for individuals with limited financial resources, leading to gaps in insurance coverage and increased risk of facing financial hardship in the event of medical emergencies.

Insurance premiums reflect the overall cost of healthcare coverage, including expenses related to medical claims, administrative costs, and profit margins for insurance companies. Rising premiums can result from factors such as increases in healthcare utilization, medical inflation, advances in medical technology, and changes in regulatory requirements.

Deductibles:

Deductibles represent the amount individuals must pay out-of-pocket for covered healthcare services before their insurance coverage begins to pay for medical expenses. Deductibles vary depending on the insurance plan and may apply to specific services, such as hospitalizations, prescriptions, or outpatient visits.

The role of deductibles is to share the cost of healthcare between individuals and insurance providers, encouraging individuals to be prudent consumers of healthcare services and avoid unnecessary medical expenditures. However, high deductibles can create financial barriers to accessing necessary medical care, particularly for individuals with chronic conditions or low-income individuals who may struggle to afford upfront costs.

Deductibles can influence healthcare utilization patterns, as individuals may delay or forgo non-urgent medical care to avoid out-of-pocket expenses. Delayed care can lead to worsening health outcomes, increased healthcare costs, and higher rates of preventable hospitalizations.

Copayments:

Copayments, also known as copays, are fixed-dollar amounts that individuals are required to pay for specific healthcare services, such as doctor visits, prescription medications, or emergency room visits. Copayments are typically paid at the time of service and serve as a cost-sharing mechanism between individuals and insurers.

Copayments play a role in controlling healthcare costs by encouraging individuals to consider the cost of services and make informed decisions about their healthcare utilization. However, excessive copayments can pose affordability barriers for individuals, particularly those with chronic conditions or low incomes, who may struggle to afford out-of-pocket expenses for necessary medical care.

65

The role of copayments is to strike a balance between ensuring access to healthcare services and controlling healthcare spending. Policymakers and insurers must consider the impact of copayments on individuals' ability to access care and design insurance plans that provide adequate financial protection for insured individuals while promoting cost-conscious healthcare utilization.

In addressing healthcare reform and improving access and affordability, policymakers, insurers, and healthcare stakeholders must carefully consider the role of insurance premiums, deductibles, and copayments in shaping individuals' access to healthcare services. Balancing the need for financial protection with the goal of controlling healthcare costs is essential in designing insurance plans and policy interventions that promote equitable access to high-quality, affordable healthcare for all Americans.

Financial Burden on Patients and Families

The financial burden on patients and families resulting from healthcare costs is a pervasive issue in the United States healthcare system. As healthcare expenses continue to rise, individuals and families face significant financial challenges that can impact their financial stability, quality of life, and overall well-being. Understanding the factors contributing to the financial burden on patients and families is essential for designing effective healthcare reform solutions aimed at improving access and affordability for all Americans.

High Insurance Premiums: The rising cost of health insurance premiums places a considerable financial strain on patients and families. Premiums represent a significant portion of household budgets, particularly for individuals and families purchasing insurance coverage through employer-sponsored plans or private insurance marketplaces. High premiums can limit individuals' ability to afford health insurance coverage, leading to gaps in coverage and increased financial vulnerability in the event of medical emergencies.

Out-of-Pocket Expenses: Out-of-pocket expenses, including deductibles, copayments, coinsurance, and prescription

drug costs, contribute to the financial burden on patients and families. Even with insurance coverage, individuals are often responsible for paying a portion of their healthcare costs, which can accumulate quickly, particularly for individuals with chronic health conditions or complex medical needs. High out-of-pocket expenses can lead to financial hardship, medical debt, and bankruptcy for patients and families struggling to afford necessary medical care.

Medical Debt and Bankruptcy:

The inability to afford healthcare expenses can result in medical debt, which affects millions of Americans each year. Medical debt arises when individuals are unable to pay for medical services and treatments, leading to unpaid medical bills and collections actions by healthcare providers and creditors. Medical debt can have long-term financial consequences, including damage to credit scores, wage garnishment, and bankruptcy, which can further exacerbate individuals' financial instability and limit their access to credit and financial resources.

Financial Trade-offs and Sacrifices:

The financial burden of healthcare costs often forces patients and families to make difficult trade-offs and sacrifices to afford medical care. Individuals may forgo necessary medical treatments, delay preventive care, or skip prescription medications to save money. Sacrificing healthcare needs can lead to adverse health outcomes, increased healthcare costs, and decreased quality of life for patients and families, perpetuating a cycle of financial insecurity and poor health.

Impact on Quality of Life and Well-being:

The financial burden of healthcare costs can take a toll on patients' and families' quality of life and overall well-being. Financial stress related to healthcare expenses can contribute to anxiety, depression, and emotional distress, affecting individuals' mental health and psychological resilience. Fear of medical debt and bankruptcy can also deter individuals from seeking necessary medical care, leading to avoidable health complications and diminished health outcomes.

69

Addressing the financial burden on patients and families requires comprehensive healthcare reform efforts aimed at improving access to affordable healthcare services and reducing healthcare costs. Policy solutions such as expanding access to affordable insurance coverage, controlling prescription drug prices, implementing price transparency measures, and increasing investment in preventive care and public health initiatives can help mitigate affordability barriers and ensure that all Americans have access to high-quality, affordable healthcare services without facing undue financial hardship. By prioritizing affordability as a key component of healthcare reform, policymakers can advance the goal of creating a more equitable, accessible, and sustainable healthcare system for all.

CHAPTER 4

POLICY APPROACHES TO REFORM

Expansion of Medicaid and Public Insurance Programs: Expanding Medicaid eligibility and increasing access to public insurance programs, such as Medicare and the Children's Health Insurance Program (CHIP), can extend coverage to uninsured and low-income individuals. Policy reforms aimed at expanding Medicaid eligibility criteria and closing the Medicaid coverage gap can help reduce disparities in access to healthcare services among vulnerable populations.

Marketplace Reforms and Insurance Regulation: Implementing marketplace reforms and strengthening insurance regulations can enhance consumer protections and promote affordability in the private insurance market. Policy approaches may include capping out-of-pocket expenses,

71

prohibiting coverage denials based on pre-existing conditions, and standardizing benefit packages to facilitate comparison shopping and improve transparency in insurance coverage options.

Price Transparency and Cost Control Measures: Enhancing price transparency and implementing cost control measures can empower consumers to make informed decisions about their healthcare spending and promote competition among healthcare providers. Policy initiatives may include requiring healthcare providers to disclose pricing information for medical services, promoting value-based reimbursement models, and implementing payment reforms aimed at reducing unnecessary utilization and incentivizing high-value care.

Prescription Drug Pricing Reforms: Addressing prescription drug pricing disparities and controlling pharmaceutical costs can help alleviate financial burdens on patients and families. Policy approaches may include negotiating drug prices with pharmaceutical manufacturers, allowing importation of lower-cost prescription drugs from

international markets, and promoting generic drug competition to increase affordability and accessibility of medications.

Investment in Primary Care and Prevention:
Increasing investment in primary care and preventive services can improve health outcomes, reduce healthcare costs, and promote population health. Policy reforms may include incentivizing healthcare providers to adopt patient-centered medical home models, expanding access to preventive screenings and immunizations, and supporting community-based initiatives aimed at addressing social determinants of health.

Health Information Technology and Care Coordination:
Enhancing health information technology infrastructure and promoting care coordination initiatives can improve care quality, reduce medical errors, and streamline healthcare delivery processes. Policy approaches may include incentivizing adoption of electronic health records, facilitating interoperability among healthcare systems, and promoting care coordination among healthcare providers

73

to improve care transitions and reduce duplicative services.

Addressing Social Determinants of Health: Recognizing the impact of social determinants of health on healthcare outcomes, policy approaches may focus on addressing underlying social and economic factors that influence health disparities. Initiatives may include investing in affordable housing programs, expanding access to nutritional assistance programs, supporting workforce development initiatives, and promoting community-based interventions aimed at addressing social determinants of health and promoting health equity.

Health Equity and Cultural Competency Training: Promoting health equity and cultural competency training among healthcare providers can improve access to culturally sensitive care and reduce disparities in healthcare delivery. Policy approaches may include integrating health equity principles into healthcare delivery systems, supporting diversity and inclusion initiatives in healthcare workforce recruitment and training, and implementing cultural

74

competency standards for healthcare providers.

By implementing comprehensive policy reforms that address the multifaceted challenges of access and affordability in the healthcare system, policymakers can work towards creating a more equitable, accessible, and sustainable healthcare system that meets the needs of all Americans. Collaborative efforts among policymakers, healthcare stakeholders, and community leaders are essential to advancing healthcare reform initiatives and improving health outcomes for individuals and families across the United States.

Review of Past Healthcare Reform Efforts:

The Affordable Care Act (ACA), signed into law in 2010, represented a landmark healthcare reform effort in the United States aimed at expanding access to affordable healthcare coverage, improving quality of care, and reducing healthcare costs. As one of the most significant legislative achievements in recent decades, the ACA introduced a range of reforms and provisions that reshaped the nation's healthcare landscape and had profound implications for patients, providers, insurers, and policymakers alike.

Expansion of Insurance Coverage:

A central objective of the ACA was to expand access to health insurance coverage for millions of uninsured Americans. The law established health insurance marketplaces where individuals and small businesses could shop for private insurance plans and access federal subsidies to offset premium costs. Medicaid expansion was another key provision of the ACA, extending eligibility criteria to cover low-income adults in participating states.

Consumer Protections and Insurance Reforms:

The ACA introduced numerous consumer protections and insurance reforms aimed at improving the quality and comprehensiveness of health insurance coverage. Provisions such as guaranteed issue, which prohibited insurers from denying coverage based on pre-existing conditions, and essential health benefits requirements ensured that individuals had access to comprehensive coverage options with essential health benefits.

Healthcare Delivery Reforms: The ACA incentivized changes in healthcare delivery and payment models aimed at promoting quality of care, care coordination, and cost containment. Initiatives such as accountable care organizations (ACOs), bundled payments, and value-based purchasing programs sought to incentivize healthcare providers to deliver high-quality, coordinated care while containing costs and improving patient outcomes.

Preventive Care and Public Health Initiatives: The ACA emphasized the importance of preventive care and public health initiatives in promoting population health and reducing healthcare costs. The law mandated coverage of preventive services, such as screenings, vaccinations, and wellness exams, without cost-sharing requirements, making preventive care more accessible and affordable for individuals and families.

Challenges and Criticisms: Despite its accomplishments, the ACA faced significant challenges and criticisms, including political opposition, legal challenges, and implementation issues.

77

Controversial aspects of the law, such as the individual mandate requiring most Americans to have health insurance or pay a penalty, generated heated debates and legal disputes. Additionally, concerns about affordability, rising premiums, and narrow provider networks raised questions about the sustainability of the law's coverage expansions.

Impact on Access and Affordability: The ACA had a profound impact on access to healthcare and affordability in the United States. The law resulted in historic reductions in the uninsured rate, expanded access to coverage for millions of Americans, and provided critical consumer protections against discriminatory insurance practices. However, affordability remained a persistent challenge for many individuals and families, particularly those who did not qualify for subsidies or faced high out-of-pocket costs.

Legacy and Lessons Learned: The ACA's legacy as a transformative piece of healthcare legislation underscores the complexity and contentiousness of healthcare reform in the United States.

While the law made significant strides in expanding access to coverage and improving consumer protections, its implementation revealed the complexities of healthcare policy and the need for ongoing efforts to address access, affordability, and quality of care in the nation's healthcare system.

As policymakers, healthcare stakeholders, and advocates continue to grapple with the challenges of access and affordability in healthcare, the ACA serves as a critical reference point and source of lessons learned for future reform efforts. By building on the successes and addressing the shortcomings of past healthcare reform initiatives, policymakers can work towards creating a more equitable, accessible, and sustainable healthcare system that meets the needs of all Americans.

Comparative Analysis of Healthcare Systems in Other Countries:

United Kingdom (National Health Service - NHS):

The NHS provides universal healthcare coverage to all UK residents, funded primarily through taxation. Healthcare services, including primary care, hospital care, and prescription medications, are provided free at the point of use.

The NHS emphasizes equitable access to healthcare services, with a focus on preventive care, health promotion, and population health management. Patients have the freedom to choose their primary care providers and specialists within the NHS system.

While the NHS offers comprehensive coverage and promotes health equity, it faces challenges such as long wait times for non-emergency procedures, funding constraints, and workforce shortages.

Canada (Single-Payer Healthcare):

Canada's healthcare system is publicly funded and administered by individual provinces and territories. The Canada Health Act ensures universal coverage for medically necessary hospital and physician services, with funding provided by the government through taxation.

Canadians have access to a wide range of healthcare services, including primary care, hospital care, and specialty services,

80

without facing financial barriers at the point of care.

However, Canada's healthcare system experiences challenges such as wait times for specialist care and elective surgeries, particularly in areas with physician shortages or limited healthcare infrastructure.

Germany (Social Health Insurance - SHI):

Germany's healthcare system is based on social health insurance, with coverage provided by multiple competing sickness funds, which are nonprofit entities funded by employer and employee contributions.

The SHI system offers comprehensive coverage for medical services, including primary care, specialist care, hospital care, and prescription medications. Patients have the freedom to choose their physicians and healthcare providers.

Germany's healthcare system emphasizes quality of care, patient satisfaction, and cost containment through competition among sickness funds and provider networks. However, the system faces challenges such as rising healthcare costs, aging population, and disparities in healthcare access among different regions.

France (Universal Health Coverage):

France's healthcare system provides universal coverage to all residents, financed through a combination of payroll taxes, government funding, and patient cost-sharing. The system is characterized by a mix of public and private providers.

Patients in France have access to comprehensive healthcare services, including primary care, hospital care, specialist services, and prescription medications. The government regulates healthcare prices and reimburses a significant portion of healthcare expenses.

France's healthcare system prioritizes accessibility, quality of care, and patient choice, with a strong emphasis on preventive care and health promotion. However, challenges such as rising healthcare expenditures, regional disparities in healthcare access, and workforce shortages persist.

Comparing healthcare systems across different countries provides valuable insights into alternative approaches to healthcare delivery, financing, and regulation. While each healthcare system has its unique strengths and challenges, lessons learned from international experiences can inform healthcare reform efforts in the United States and contribute to the development of more equitable, accessible, and affordable healthcare solutions.

Proposed Policy Solutions for Improving Access and Affordability

Expansion of Medicaid Eligibility:

Proposal: Expand Medicaid eligibility criteria to cover more low-income individuals and families, particularly in states that have not yet expanded Medicaid under the Affordable Care Act (ACA).

Rationale: Medicaid expansion has been shown to significantly increase access to healthcare services and reduce uninsured rates among vulnerable populations. By expanding eligibility, more individuals will have access to

comprehensive coverage, preventive care, and essential health services, leading to improved health outcomes and reduced financial barriers to care.

Enhancement of Marketplace Subsidies:

Proposal: Enhance premium subsidies and cost-sharing reductions for individuals and families purchasing insurance coverage through the health insurance marketplaces established under the ACA.

Rationale: Increasing subsidies for lower- and middle-income individuals can help make insurance premiums more affordable and reduce out-of-pocket expenses for healthcare services. Enhanced subsidies will make coverage more accessible and affordable for individuals and families, particularly those with limited financial resources or high healthcare needs.

Implementation of a Public Option:

Proposal: Introduce a public health insurance option that competes with private insurance plans in the marketplace, offering individuals and employers an alternative choice for coverage.

Rationale: A public option can increase competition in the insurance market, driving down premiums and expanding access to affordable coverage options. By providing a government-administered alternative to private insurance, a public option can promote choice, transparency, and affordability in healthcare coverage.

Regulation of Prescription Drug Prices:

Proposal: Implement measures to regulate prescription drug prices, including negotiation of drug prices, importation of lower-cost medications, and capping price increases for essential medications.

Rationale: Prescription drug costs represent a significant financial burden for patients and families, contributing to affordability barriers and medication non-adherence. Regulating drug prices can help lower costs for consumers, improve medication affordability, and promote access to essential treatments and therapies.

Investment in Telehealth and Telemedicine:

Proposal: Expand access to telehealth and telemedicine services by investing in infrastructure, technology, and reimbursement mechanisms that support virtual care delivery.

Rationale: Telehealth has the potential to improve access to healthcare services, particularly in rural and underserved areas where access to providers may be limited. By leveraging technology to deliver care remotely, patients can access timely and convenient medical services while reducing travel time and costs associated with in-person visits.

Addressing Social Determinants of Health:

Proposal: Invest in initiatives that address social determinants of health, such as affordable housing programs, nutrition assistance, job training, and community health programs.

Rationale: Social determinants of health, including income, education, housing, and access to healthy food, significantly influence individuals' health outcomes and healthcare utilization. By addressing underlying social and economic factors, policymakers can improve population health, reduce healthcare disparities, and promote health equity across diverse communities.

Support for Value-Based Care and Payment Reform:

Proposal: Promote value-based care delivery models and payment reforms that incentivize high-quality, coordinated care, improve health outcomes, and contain healthcare costs.

Rationale: Value-based care models focus on improving patient outcomes and reducing healthcare expenditures through care coordination, preventive care, and evidence-based practices. By shifting from fee-for-service reimbursement to value-based payment models, healthcare providers are incentivized to prioritize quality, efficiency, and patient-centered care.

By implementing these proposed policy solutions and prioritizing access and affordability in healthcare reform efforts, policymakers can work towards creating a more equitable, accessible, and sustainable healthcare system that meets the needs of all Americans. Collaboration among policymakers, healthcare stakeholders, and community leaders is

essential to advancing meaningful reforms
that address the root causes of healthcare
disparities and ensure that healthcare is
accessible and affordable for individuals
and families across the United States.

CHAPTER 5
INNOVATION IN
HEALTHCARE DELIVERY

Telehealth and Telemedicine:

Telehealth and telemedicine technologies enable healthcare providers to deliver medical services remotely, expanding access to care for patients in rural and underserved areas. Through virtual consultations, remote monitoring, and telehealth platforms, patients can access timely medical advice, receive chronic disease management, and participate in remote diagnostic testing without the need for in-person visits.

Telehealth innovations have the potential to improve healthcare access, reduce healthcare costs, and enhance patient convenience. By leveraging telehealth technologies, healthcare providers can streamline care delivery, improve care coordination, and reach patients in remote or geographically isolated communities.

Mobile Health (mHealth) Applications:

Mobile health applications, or mHealth apps, empower patients to take control of their health through digital tools and resources accessible via smartphones and wearable devices. From medication reminders and symptom tracking to virtual coaching and wellness programs, mHealth apps enable individuals to manage chronic conditions, monitor health metrics, and engage in preventive care activities from the palm of their hand.

mHealth innovations promote patient engagement, self-management, and adherence to treatment plans, leading to better health outcomes and reduced healthcare utilization. By leveraging mobile technologies, healthcare providers can personalize care experiences, deliver targeted interventions, and empower patients to make informed decisions about their health and well-being.

Remote Patient Monitoring (RPM):

Remote patient monitoring (RPM) solutions allow healthcare providers to monitor patients' vital signs, health metrics, and

90

disease progression from a distance, enabling proactive interventions and timely adjustments to treatment plans. Through wearable sensors, connected devices, and remote monitoring platforms, patients can transmit real-time health data to their care teams, facilitating continuous monitoring and early detection of health issues.

RPM innovations enhance care coordination, improve patient outcomes, and reduce hospital readmissions by enabling timely interventions and preventive measures. By remotely monitoring patients' health status and adherence to treatment protocols, healthcare providers can optimize care delivery, minimize complications, and support patients in managing chronic conditions from home.

Precision Medicine and Genomics:

Precision medicine and genomics revolutionize healthcare delivery by tailoring treatment approaches to individual patients' unique genetic makeup, health profiles, and disease characteristics. By leveraging advances in genomic sequencing, biomarker analysis, and personalized diagnostics, healthcare providers can identify genetic predispositions, predict treatment responses, and optimize therapeutic interventions for patients.

Precision medicine innovations empower healthcare providers to deliver targeted, evidence-based treatments that are tailored to each patient's specific needs and genetic profile. By integrating genomic data into clinical decision-making, healthcare providers can optimize treatment efficacy, minimize adverse effects, and improve patient outcomes across a wide range of medical conditions.

Integrated Care Models:

Integrated care models promote collaboration and coordination among multidisciplinary healthcare teams, including primary care providers, specialists, social workers, and community organizations. By integrating medical, behavioral, and social services into cohesive care delivery systems, integrated care models address patients' holistic needs, improve care continuity, and enhance care quality.

Integrated care innovations support care coordination, reduce fragmentation, and

92

improve patient outcomes by facilitating seamless transitions across care settings and aligning resources to meet patients' comprehensive needs. By fostering interdisciplinary collaboration and communication, integrated care models promote patient-centered care experiences and support individuals in achieving optimal health and well-being.

Innovation in healthcare delivery holds tremendous promise for transforming the way healthcare services are accessed, delivered, and experienced by patients and providers alike. By embracing innovative technologies, care models, and approaches to care delivery, healthcare systems can drive improvements in access, affordability, and quality of care, ultimately advancing the goal of creating a more equitable, accessible, and patient-centered healthcare system for all.

Technological Advancements and Their Potential to Reduce Costs

Artificial Intelligence (AI) and Machine Learning:

AI and machine learning technologies have the potential to streamline administrative processes, automate repetitive tasks, and optimize resource allocation in healthcare organizations. By leveraging AI-powered algorithms, healthcare providers can improve operational efficiency, reduce administrative burdens, and minimize unnecessary healthcare expenditures associated with manual workflows and inefficiencies.

Predictive Analytics and Data Analytics:

Predictive analytics and data analytics tools enable healthcare organizations to analyze large volumes of clinical and administrative data to identify patterns, trends, and risk factors associated with patient outcomes, healthcare utilization, and cost drivers. By leveraging predictive analytics algorithms, healthcare providers

94

can identify high-risk patients, intervene early to prevent complications, and optimize resource allocation to improve care outcomes and reduce avoidable costs.

Telehealth and Remote Monitoring Technologies:

Telehealth and remote monitoring technologies enable healthcare providers to deliver virtual care services, monitor patients remotely, and facilitate remote consultations and follow-up visits. By leveraging telehealth platforms and remote monitoring devices, healthcare organizations can reduce unnecessary hospitalizations, emergency department visits, and outpatient appointments, leading to cost savings associated with reduced healthcare utilization and improved care coordination.

Health Information Exchange (HIE) and Interoperability Solutions:

Health information exchange (HIE) and interoperability solutions enable seamless exchange of patient health information across disparate healthcare systems, providers, and care settings. By promoting interoperability and data sharing, HIE technologies facilitate care coordination, reduce duplication of services, and improve care transitions, leading to cost savings associated with reduced medical errors, redundant testing, and administrative inefficiencies.

Robotics and Automation in Healthcare Delivery:

Robotics and automation technologies have the potential to revolutionize healthcare delivery by automating routine tasks, assisting healthcare providers in surgical procedures, and enhancing precision and efficiency in clinical workflows. By deploying robotics and automation solutions, healthcare organizations can improve patient safety, reduce surgical complications, and optimize resource utilization, leading to cost savings associated with improved clinical outcomes and operational efficiency.

Digital Health Platforms and Mobile Applications:

Digital health platforms and mobile applications empower patients to actively engage in their healthcare management, access health information, and participate in virtual care services. By leveraging digital health solutions, healthcare organizations can promote patient self-management, facilitate remote monitoring, and enhance patient-provider communication, leading to cost savings associated with improved patient adherence, reduced hospital readmissions, and enhanced patient satisfaction.

Precision Medicine and Personalized Therapies:

Precision medicine and personalized therapies leverage advances in genomic sequencing, molecular diagnostics, and targeted treatments to tailor medical interventions to individual patient characteristics, genetic profiles, and disease pathways. By embracing precision medicine approaches, healthcare providers can optimize treatment efficacy, minimize adverse effects, and improve patient outcomes, leading to cost savings associated with reduced treatment failures, adverse events, and unnecessary healthcare expenditures.

Technological advancements have the potential to revolutionize healthcare

97

delivery, improve patient outcomes, and reduce costs across the healthcare continuum. By embracing innovation and leveraging technology-driven solutions, healthcare organizations can enhance operational efficiency, optimize resource utilization, and promote value-based care delivery, ultimately advancing the goal of creating a more accessible, affordable, and sustainable healthcare system for all.

Telemedicine and Its Role in Increasing Access

Telemedicine refers to the remote delivery of healthcare services using telecommunications technology, such as video conferencing, mobile apps, and secure messaging platforms. It has emerged as a powerful tool in healthcare delivery, playing a crucial role in increasing access to medical care, particularly in underserved and remote areas. Here's how telemedicine contributes to improving access as a solution for **healthcare reform:**

Overcoming Geographic Barriers: Telemedicine enables patients to connect with healthcare providers

98

regardless of their geographic location. In rural and remote areas where access to healthcare facilities may be limited, telemedicine bridges the gap by allowing patients to consult with specialists and primary care providers remotely, reducing the need for travel and overcoming geographic barriers to care.

Expanding Specialty Care Access:
Telemedicine expands access to specialty care services, such as cardiology, dermatology, and mental health, by connecting patients with specialists who may not be available locally. Through telemedicine platforms, patients can access expert consultations, receive timely diagnoses, and access specialized treatments without the need for lengthy referral processes or travel to distant healthcare facilities.

Improving Timely Access to Care:
Telemedicine enhances timely access to medical care by offering flexible appointment scheduling, extended hours of operation, and on-demand consultations. Patients can access healthcare services when they need them, reducing wait times for appointments, avoiding unnecessary

99

emergency room visits, and receiving timely interventions for acute and chronic health conditions.

Enhancing Care Coordination and Follow-Up: Telemedicine supports care coordination and follow-up by facilitating communication between healthcare providers, patients, and caregivers. Through virtual visits and remote monitoring, healthcare teams can collaborate on care plans, review treatment progress, and provide ongoing support to patients managing chronic conditions, promoting continuity of care and reducing gaps in care delivery.

Increasing Patient Engagement and Education: Telemedicine promotes patient engagement and education by empowering individuals to actively participate in their healthcare management. Through telehealth platforms and mobile apps, patients can access educational resources, track health metrics, and communicate with their healthcare providers, fostering self-management skills and promoting preventive care behaviors.

100

Reducing Healthcare Costs:

Telemedicine has the potential to reduce healthcare costs by eliminating unnecessary healthcare utilization, such as non-urgent emergency room visits and avoidable hospital readmissions. By providing timely interventions, preventive care services, and remote monitoring, telemedicine helps optimize resource utilization, reduce healthcare expenditures, and improve cost-effectiveness in care delivery.

Addressing Healthcare Disparities:

Telemedicine addresses healthcare disparities by expanding access to care for underserved populations, including rural communities, low-income individuals, and those with limited mobility or transportation barriers. By leveraging telemedicine solutions, healthcare organizations can reach vulnerable populations, improve health outcomes, and promote health equity across diverse communities.

In conclusion, telemedicine plays a vital role in increasing access to healthcare services, improving patient outcomes, and enhancing affordability in the United States

101

healthcare system. By leveraging telemedicine technologies and integrating telehealth solutions into care delivery models, policymakers, healthcare providers, and stakeholders can advance the goal of creating a more accessible, equitable, and patient-centered healthcare system that meets the needs of all Americans.

Value-Based Care Models and Preventive Healthcare Initiatives

Value-Based Care Models:

Value-based care models shift the focus of healthcare delivery from volume-based reimbursement to value-based reimbursement, prioritizing quality of care, patient outcomes, and cost-effectiveness. These models incentivize healthcare providers to deliver high-quality, coordinated care that emphasizes prevention, early intervention, and patient-centeredness.

Accountable Care Organizations (ACOs): ACOs are groups of healthcare providers and organizations that

collaborate to improve care coordination, manage population health, and achieve shared savings through value-based contracts. ACOs focus on promoting preventive care, reducing hospital admissions, and optimizing resource utilization to improve patient outcomes and reduce healthcare costs.

Bundled Payment Models: Bundled payment models reimburse healthcare providers for a bundle of services related to a specific episode of care, such as a surgical procedure or chronic disease management. By aligning incentives and accountability across the care continuum, bundled payment models encourage care coordination, reduce fragmentation, and promote efficiency in care delivery.

Patient-Centered Medical Homes (PCMHs):

PCMHs are primary care practices that serve as central hubs for coordinating patient care, managing chronic conditions, and promoting preventive care services. PCMHs emphasize team-based care, care coordination, and patient engagement to improve health outcomes, enhance patient experiences, and reduce healthcare costs through proactive management of chronic diseases and preventive interventions.

Preventive Healthcare Initiatives:

Preventive healthcare initiatives focus on promoting health and wellness, preventing disease, and addressing risk factors before they escalate into serious health conditions. These initiatives emphasize early detection, health education, and lifestyle modifications to reduce the burden of chronic diseases and improve overall population health.

Immunization Programs:

Immunization programs promote vaccination against preventable diseases, such as influenza, measles, and HPV, to reduce the incidence of infectious diseases and protect public health. By increasing immunization rates, healthcare organizations can prevent outbreaks, reduce healthcare expenditures, and improve population health outcomes.

Screening and Early Detection:

Screening programs aim to detect health conditions in their early stages when interventions are most effective. Common screening tests include mammograms for breast cancer, colonoscopies for colorectal cancer, and blood pressure screenings for hypertension. By promoting regular screenings and early detection, healthcare providers can identify health risks, initiate timely interventions, and improve health outcomes for individuals.

Lifestyle Modification Programs:

Lifestyle modification programs encourage individuals to adopt healthy behaviors, such as regular exercise, balanced nutrition, smoking cessation, and stress management, to reduce the risk of chronic diseases and improve overall well-being. By promoting healthy lifestyle choices, healthcare organizations can prevent the onset of chronic conditions, reduce healthcare costs, and promote long-term health and vitality.

Health Education and Promotion:

Health education and promotion initiatives provide individuals and communities with information, resources, and support to make informed decisions about their health and well-being. These initiatives focus on raising awareness about healthy lifestyle choices, disease prevention strategies, and access to healthcare services, empowering

105

individuals to take control of their health and become active participants in their healthcare management.

By implementing value-based care models and preventive healthcare initiatives, policymakers, healthcare providers, and stakeholders can promote access to high-quality, affordable healthcare services, improve health outcomes, and enhance the overall value of the healthcare system. These solutions emphasize proactive management of health risks, early intervention, and patient-centered care approaches to address the underlying drivers of healthcare costs and improve population health outcomes.

CHAPTER 6

ADDRESSING SOCIAL DETERMINANTS OF HEALTH

Affordable Housing Initiatives:

Proposal: Implement affordable housing programs and initiatives to address housing insecurity and homelessness among vulnerable populations. These programs may include subsidies for low-income individuals and families, supportive housing services for individuals with chronic health conditions, and initiatives to increase affordable housing stock in high-cost areas.

Rationale: **Access to stable and affordable housing is a fundamental determinant of health, impacting individuals' ability to maintain physical and mental well-being. By addressing housing insecurity and homelessness, healthcare organizations can reduce healthcare utilization, improve health outcomes, and promote health equity among underserved populations.**

Nutrition Assistance Programs:

Proposal: Expand access to nutrition assistance programs, such as the Supplemental Nutrition Assistance Program (SNAP), school meal programs, and community food banks, to address food insecurity and promote healthy eating habits among low-income individuals and families.

Rationale: Adequate nutrition plays a crucial role in maintaining optimal health and preventing chronic diseases. By providing access to nutritious foods and nutrition education, healthcare organizations can improve dietary habits, reduce the risk of nutrition-related health conditions, and enhance overall well-being among vulnerable populations.

Employment and Economic Development Initiatives:

Proposal: Support workforce development programs, job training initiatives, and economic development initiatives aimed at reducing unemployment, poverty, and income inequality in underserved communities.

Rationale: Employment and economic stability are key determinants of health,

108

influencing individuals' access to healthcare services, social support networks, and resources for health promotion. By promoting economic opportunity and financial security, healthcare organizations can address underlying social determinants of health and improve health outcomes for individuals and families.

Access to Education and Literacy Programs:

Proposal: Invest in education and literacy programs that promote early childhood development, school readiness, and adult literacy skills, particularly among disadvantaged populations.

Rationale: Education and literacy levels have a profound impact on health outcomes, influencing individuals' ability to access healthcare information, navigate healthcare systems, and make informed decisions about their health. By promoting educational attainment and literacy skills, healthcare organizations can empower individuals to advocate for their health needs, engage in preventive care, and effectively manage chronic conditions.

Community-Based Health Promotion and Wellness Initiatives:

Proposal: Support community-based health promotion and wellness initiatives that address social isolation, mental health stigma, and barriers to physical activity and recreation in underserved communities.

Rationale:
Social connectedness, mental well-being, and physical activity are critical determinants of health, contributing to individuals' overall quality of life and resilience to health challenges. By fostering supportive social environments, promoting mental health awareness, and providing opportunities for physical activity and recreation, healthcare organizations can promote holistic approaches to health promotion and wellness.

Collaborative Partnerships and Cross-Sectoral Collaboration:

Proposal:
Foster collaborative partnerships and cross-sectoral collaboration among healthcare organizations, social service agencies, community-based organizations, and

110

government entities to address social determinants of health comprehensively.

Rationale: Addressing social determinants of health requires a multifaceted and collaborative approach that engages stakeholders across various sectors, including healthcare, housing, education, employment, and social services. By leveraging collective expertise, resources, and partnerships, healthcare organizations can develop holistic solutions that address the root causes of health disparities and promote health equity for all individuals and communities.

By addressing social determinants of health through targeted interventions, collaborative partnerships, and systemic reforms, healthcare organizations can advance the goal of creating a more equitable, accessible, and sustainable healthcare system that promotes the health and well-being of all individuals and communities.

Impact of Social Factors on Healthcare Outcomes

Socioeconomic Status (SES):

Lower socioeconomic status is associated with poorer healthcare outcomes, including higher rates of chronic diseases, lower life expectancy, and increased healthcare utilization for preventable conditions.

Individuals with low SES often face barriers to accessing healthcare services, including financial constraints, limited health literacy, and transportation challenges, which can contribute to delayed diagnosis, inadequate treatment, and poorer health outcomes.

Education Level:

Education level influences healthcare outcomes by shaping individuals' health behaviors, healthcare utilization patterns, and access to health information and resources.

Higher education levels are associated with healthier lifestyles, better health knowledge, and greater adherence to preventive care measures, leading to reduced morbidity and mortality rates and improved overall health outcomes.

Race and Ethnicity:

Racial and ethnic disparities persist in healthcare outcomes, with minority populations experiencing higher rates of chronic diseases, lower quality of care, and increased mortality rates compared to non-Hispanic white individuals.

Structural racism, discrimination, cultural differences, and language barriers contribute to disparities in healthcare access, treatment options, and health outcomes among racial and ethnic minority groups.

Housing and Neighborhood Environment:

Housing conditions and neighborhood environment influence healthcare outcomes by impacting exposure to environmental toxins, access to healthy food options, and opportunities for physical activity and recreation.

Individuals living in substandard housing or disadvantaged neighborhoods are at higher risk for health problems, including respiratory conditions, cardiovascular diseases, and mental health disorders, due to exposure to environmental hazards and lack of access to healthcare services.

Social Support Networks:

Social support networks play a crucial role in healthcare outcomes by providing emotional support, practical assistance,

113

and encouragement for healthy behaviors and healthcare management.

Individuals with strong social support networks tend to experience better health outcomes, faster recovery from illness, and improved quality of life compared to those who lack social support, highlighting the importance of social connections in promoting health and well-being.

Access to Healthcare Services:

Social factors such as income, education, and race/ethnicity influence access to healthcare services, including primary care, specialty care, and preventive services.

Individuals from disadvantaged backgrounds are more likely to experience barriers to accessing healthcare, including lack of health insurance, transportation challenges, and limited availability of healthcare providers in underserved areas, which can result in delayed diagnosis, unmet healthcare needs, and poorer health outcomes.

Understanding the impact of social factors on healthcare outcomes is essential for designing effective interventions and policies aimed at addressing health disparities, promoting health equity, and

114

improving healthcare access and affordability for all individuals and communities. By addressing social determinants of health comprehensively and adopting a holistic approach to healthcare reform, policymakers, healthcare providers, and stakeholders can work towards creating a more equitable, accessible, and patient-centered healthcare system that meets the needs of diverse populations.

Community-Based Interventions to Address Disparities

Community Health Outreach Programs:

Implement community health outreach programs that provide education, screenings, and preventive services to underserved populations. These programs can be conducted in partnership with local community organizations, churches, schools, and healthcare providers to reach individuals who may face barriers to accessing traditional healthcare services.

Mobile Health Clinics:

Deploy mobile health clinics to bring healthcare services directly to underserved communities, including rural areas, low-income neighborhoods, and areas with limited access to healthcare facilities. Mobile clinics can offer primary care, preventive screenings, vaccinations, and chronic disease management services, improving access to care for individuals who may face transportation barriers or lack nearby healthcare options.

Health Literacy and Education Campaigns:

Launch health literacy and education campaigns aimed at empowering individuals and families with the knowledge and skills to make informed healthcare decisions, navigate the healthcare system, and adopt healthy behaviors. These campaigns can include workshops, seminars, educational materials, and digital resources tailored to the cultural and linguistic needs of diverse communities.

116

Community Health Worker Programs:

Establish community health worker programs to provide culturally competent outreach, advocacy, and support services to underserved populations. Community health workers, who are trusted members of the community, can serve as liaisons between healthcare providers and patients, helping to bridge language, cultural, and socioeconomic barriers to care.

Access to Healthy Food Initiatives:

Implement initiatives to improve access to healthy and affordable food options in underserved communities, such as food deserts and low-income neighborhoods. These initiatives may include community gardens, farmers' markets, food cooperatives, and nutrition assistance programs designed to promote healthy eating habits and reduce food insecurity.

Culturally Tailored Health Programs:

Develop culturally tailored health programs and interventions that acknowledge and respect the cultural beliefs, practices, and

117

preferences of diverse populations. By incorporating cultural sensitivity and cultural competence into healthcare delivery, organizations can build trust, enhance engagement, and improve health outcomes among underserved communities.

Community Health Needs Assessments:

Conduct community health needs assessments to identify priority areas for intervention and allocation of resources. By engaging community members, stakeholders, and healthcare providers in the assessment process, organizations can gain valuable insights into local health challenges, disparities, and unmet needs, informing targeted interventions and strategies for addressing health inequities.

Collaborative Partnerships and Coalitions:

Foster collaborative partnerships and coalitions among healthcare organizations, community-based organizations, government agencies, and local stakeholders to address health disparities comprehensively. By pooling resources, expertise, and collective efforts,

118

organizations can leverage strengths, share best practices, and implement sustainable solutions to improve access to care and health outcomes for underserved populations.

By implementing community-based interventions that address the social determinants of health, promote health equity, and empower individuals and communities, healthcare organizations and stakeholders can work towards reducing disparities, improving access to care, and advancing the goal of creating a more equitable and inclusive healthcare system for all.

Collaborative Efforts Between Healthcare Providers and Social Service Organizations

Care Coordination and Case Management:

Healthcare providers and social service organizations can collaborate to facilitate care coordination and case management for individuals with complex medical and social needs. By sharing information, resources, and expertise, care teams can develop holistic care plans that address both medical and social determinants of health, ensuring comprehensive support for patients.

Screening and Referral Programs:

Healthcare providers can integrate social determinants of health screening tools into clinical assessments to identify patients' unmet social needs, such as housing instability, food insecurity, and transportation barriers. Social service organizations can then provide targeted interventions, referrals, and support services to address these needs, improving patients' overall health and well-being.

Integrated Service Delivery Models:

Implement integrated service delivery models that bring together healthcare providers, social workers, counselors, and community health workers to provide comprehensive care in a coordinated and collaborative manner. These models may include co-located services, multidisciplinary care teams, and shared electronic health records to streamline communication and care coordination across healthcare and social service settings.

Cross-Sectoral Partnerships and Coalitions:

Foster cross-sectoral partnerships and coalitions between healthcare providers, social service organizations, government agencies, philanthropic groups, and community stakeholders to address systemic issues and root causes of health disparities. By working together, stakeholders can leverage collective resources, advocate for policy changes, and implement community-wide initiatives to promote health equity and social justice.

Training and Capacity Building:

Provide training and capacity-building opportunities for healthcare providers and social service professionals to enhance their knowledge, skills, and competencies in addressing social determinants of health. Training programs may focus on cultural competency, trauma-informed care, motivational interviewing, and community engagement strategies to improve collaboration and communication among interdisciplinary teams.

121

Data Sharing and Analytics:

Establish mechanisms for data sharing and analytics to track health outcomes, identify trends, and measure the impact of collaborative efforts on population health and healthcare utilization. By sharing data insights and outcomes data, healthcare providers and social service organizations can assess program effectiveness, identify areas for improvement, and refine intervention strategies to better meet the needs of vulnerable populations.

Policy Advocacy and Systems Change:

Advocate for policy changes and systems reforms that address social determinants of health, promote health equity, and support the integration of healthcare and social services. Healthcare providers and social service organizations can collaborate to advocate for funding, resources, and policy initiatives that prioritize prevention, early intervention, and community-based approaches to improving access and affordability of care.

By fostering collaborative partnerships between healthcare providers and social service organizations, stakeholders can work together to address the complex interplay of medical, social, and economic factors that influence health outcomes and disparities. Through shared goals, shared resources, and shared accountability, collaborative efforts can lead to meaningful improvements in access to care, health outcomes, and overall well-being for individuals and communities across the United States.

CHAPTER 7

EMPOWERING PATIENTS AND PROVIDERS

Patient Education and Health Literacy:

Provide comprehensive patient education and health literacy programs to empower individuals with the knowledge and skills to make informed healthcare decisions, understand their rights and responsibilities, and actively participate in their care.

Offer accessible and culturally sensitive educational materials, workshops, and resources that address common health concerns, promote preventive care practices, and encourage active engagement in managing chronic conditions.

Shared Decision-Making:

Promote shared decision-making between patients and healthcare providers by facilitating open communication, mutual respect, and collaboration in healthcare decision-making processes.

124

Encourage healthcare providers to engage patients in discussions about treatment options, risks, benefits, and preferences, allowing patients to play an active role in determining their care plans based on their values, goals, and personal circumstances.

Access to Health Information and Technology:

Expand access to health information and technology platforms that enable patients to securely access their medical records, track health metrics, and communicate with their healthcare providers.

Encourage the adoption of patient portals, telehealth services, mobile health apps, and other digital tools that empower patients to manage their health, schedule appointments, request prescription refills, and receive personalized health recommendations.

Patient Advocacy and Support Groups:

Foster patient advocacy and support groups that provide individuals with peer support, encouragement, and resources to navigate healthcare systems, access needed services, and advocate for their health needs.

125

Facilitate the formation of patient-led organizations, online communities, and support networks that offer emotional support, practical guidance, and advocacy opportunities for patients and caregivers facing similar health challenges.

Provider Training and Continuing Education:

Offer provider training and continuing education programs that emphasize patient-centered care principles, communication skills, cultural competence, and empathy in healthcare delivery.

Provide healthcare providers with opportunities to enhance their knowledge, competencies, and understanding of social determinants of health, health disparities, and strategies for addressing patients' diverse needs and preferences.

Quality Improvement and Feedback Mechanisms:

Implement quality improvement initiatives and feedback mechanisms that solicit patient input, experiences, and perspectives to inform healthcare delivery improvements, service redesign, and organizational policies.

126

Encourage patients to provide feedback on their healthcare experiences, satisfaction levels, and suggestions for enhancing care delivery through surveys, focus groups, patient advisory councils, and online feedback platforms.

Collaborative Care Models:

Foster collaborative care models that promote interdisciplinary teamwork, care coordination, and integration of services across healthcare settings, specialties, and disciplines.

Support care teams in working together to address patients' holistic needs, enhance care continuity, and optimize outcomes through shared decision-making, care planning, and communication.

By empowering both patients and providers with the knowledge, skills, resources, and support needed to actively engage in healthcare decision-making, communication, and collaboration, stakeholders can foster a patient-centered healthcare system that prioritizes individual needs, preferences, and outcomes. Through ongoing efforts to promote empowerment and partnership between patients and providers, healthcare reform can drive meaningful improvements

127

in access, affordability, and quality of care for all individuals and communities across the United States.

Patient Education and Advocacy Initiatives

Health Literacy Programs:

Develop and implement health literacy programs that empower patients with the knowledge and skills to understand healthcare information, navigate healthcare systems, and make informed decisions about their health and treatment options.

Offer workshops, seminars, and educational materials that cover topics such as understanding medical terminology, interpreting lab results, medication management, and preventive care practices.

Disease-specific Education Campaigns:

Launch disease-specific education campaigns that provide patients with information, resources, and support related to managing chronic conditions, such as diabetes, heart disease, asthma, and cancer.

Collaborate with healthcare providers, patient advocacy organizations, and community groups to raise awareness, promote early detection, and encourage adherence to treatment regimens among individuals affected by common health conditions.

Patient Navigation Services:

Establish patient navigation services that offer personalized assistance and support to help patients navigate the complexities of the healthcare system, overcome barriers to care, and access needed services.

Train patient navigators to provide advocacy, guidance, and emotional support to patients and their families throughout the healthcare journey, from scheduling appointments and coordinating referrals to understanding insurance coverage and accessing community resources.

Empowerment Workshops and Support Groups:

Facilitate empowerment workshops and support groups that provide patients with opportunities to share experiences, learn from peers, and develop self-management

129

skills for coping with chronic illnesses, disabilities, or caregiving responsibilities.

Offer support group meetings, wellness activities, and peer mentoring programs that promote resilience, peer support, and empowerment among individuals facing similar health challenges or life transitions.

Healthcare Rights and Advocacy Training:

Provide healthcare rights and advocacy training to equip patients with knowledge of their rights, responsibilities, and legal protections in healthcare settings.

Educate patients about their options for lodging complaints, appealing denials, and seeking assistance from patient advocacy organizations, ombudsman programs, and regulatory agencies to address grievances and resolve disputes related to healthcare services.

Digital Health Tools and Resources:

Leverage digital health tools and resources to enhance patient education, engagement, and empowerment through online platforms, mobile apps, and telehealth services.

Develop interactive educational materials, self-assessment tools, and decision aids that empower patients to actively participate in healthcare decision-making, track health metrics, and communicate with their healthcare providers in real-time.

Community Health Fairs and Outreach Events:

Organize community health fairs, outreach events, and wellness screenings to engage patients and families in preventive health measures, health promotion activities, and access to healthcare services.

Partner with local healthcare providers, community organizations, and volunteers to offer free screenings, health assessments, immunizations, and health education workshops in underserved communities.

By investing in patient education and advocacy initiatives, healthcare organizations, policymakers, and stakeholders can empower individuals to take charge of their health, make informed decisions about their care, and advocate for policies and practices that promote access, affordability, and quality in healthcare delivery. Through collaborative efforts to strengthen patient education and advocacy efforts, healthcare reform can drive meaningful improvements in health outcomes, patient satisfaction, and overall well-being for individuals and communities across the United States.

Support for Healthcare Providers in Delivering Quality Care

Continuing Education and Training Programs:

Offer continuing education and training programs for healthcare providers to stay updated on best practices, evidence-based guidelines, and advancements in medical technology and treatment modalities.

Provide opportunities for professional development, skill enhancement, and interdisciplinary collaboration to support healthcare providers in delivering high-quality, patient-centered care.

Clinical Decision Support Tools:

Implement clinical decision support tools, electronic health records (EHRs), and point-of-care resources that assist healthcare providers in making evidence-based decisions, diagnosing conditions, and managing treatment plans.

Integrate decision support systems that provide real-time alerts, reminders, and clinical guidelines to improve adherence to clinical protocols, reduce medical errors, and enhance patient safety.

Team-Based Care Models:

Adopt team-based care models that leverage the expertise of diverse healthcare professionals, including physicians, nurses, pharmacists, social workers, and care coordinators, to deliver comprehensive and coordinated care.

Foster a collaborative practice environment that promotes effective communication, shared decision-making, and mutual respect among team members, enhancing care coordination and optimizing patient outcomes.

Health Information Technology Infrastructure:

Invest in robust health information technology (HIT) infrastructure, interoperable systems, and data analytics capabilities to streamline clinical workflows, facilitate information sharing,

133

and improve care coordination across healthcare settings.

Implement user-friendly EHR platforms, telemedicine solutions, and mobile health applications that support seamless communication, documentation, and access to patient information for healthcare providers.

Quality Improvement Initiatives:

Establish quality improvement initiatives, performance metrics, and clinical outcome measures to monitor and evaluate the effectiveness of healthcare delivery processes, identify areas for improvement, and drive continuous quality improvement efforts.

Engage healthcare providers in quality improvement projects, peer review activities, and interdisciplinary committees to promote a culture of accountability, transparency, and excellence in clinical practice.

Workforce Well-Being and Resilience Programs:

Prioritize workforce well-being and resilience by implementing programs and resources that address burnout, compassion fatigue, and stress among healthcare providers.

Offer wellness initiatives, mindfulness training, mental health support services, and peer support networks to help healthcare providers cope with the demands of clinical practice and maintain a healthy work-life balance.

Patient-Centered Care Principles:

Emphasize patient-centered care principles that prioritize patients' preferences, values, and goals in treatment decisions, care planning, and care delivery processes.

Encourage healthcare providers to engage patients in meaningful discussions, actively listen to their concerns, and involve them as partners in their healthcare journey to enhance patient satisfaction and improve health outcomes.

By providing comprehensive support systems, resources, and infrastructure for healthcare providers, healthcare reform efforts can empower providers to deliver high-quality, patient-centered care that improves access, affordability, and effectiveness in healthcare delivery. Through collaborative efforts to address the needs of healthcare providers, stakeholders can create a supportive environment that fosters excellence in clinical practice and enhances the overall quality of care for individuals and communities across the United States.

Strategies for Promoting Shared Decision-Making and Patient-Centered Care

Patient Education and Empowerment:

Provide comprehensive patient education materials and resources that empower individuals to actively participate in their healthcare decisions.

Offer information about treatment options, risks, benefits, and alternatives in plain language, using visual aids and multimedia formats to enhance understanding and engagement.

Shared Decision-Making Tools and Decision Aids:

Integrate shared decision-making tools and decision aids into clinical practice to facilitate discussions between healthcare providers and patients about treatment options and preferences.

Utilize decision aids such as videos, pamphlets, and online platforms to present information about different treatment options, outcomes, and potential side effects in a structured and unbiased manner.

Patient Decision Support Resources:

Offer patient decision support resources, including online portals, decision-making guides, and interactive tools that enable individuals to explore their values, preferences, and treatment priorities.

Provide access to decision support resources that help patients weigh the pros and cons of different treatment options, clarify their goals of care, and make informed decisions that align with their personal values and preferences.

Provider Training in Communication Skills:

Provide training and support for healthcare providers in communication skills, active listening, and shared decision-making techniques.

Offer workshops, role-playing exercises, and continuing education programs that enhance providers' ability to engage patients in meaningful conversations, elicit their perspectives, and address their concerns and preferences.

Emphasis on Patient Preferences and Goals:

Prioritize patient preferences, values, and goals in treatment decisions, care planning, and goal-setting processes.

Encourage healthcare providers to explore patients' individual preferences, cultural beliefs, and social contexts to tailor care plans that reflect patients' unique needs and priorities.

Encouragement of Family and Caregiver Involvement:

Involve family members, caregivers, and support networks in shared decision-making processes to ensure that patients receive holistic and culturally responsive care.

Facilitate open communication and collaboration among patients, families, and healthcare providers to address shared concerns, preferences, and goals of care.

Promotion of Health Literacy and Informed Consent:

Promote health literacy and informed consent practices that enable patients to understand their treatment options, participate in decision-making processes, and provide informed consent for medical interventions.

Provide clear and accessible information about risks, benefits, and alternatives to empower patients to make choices that align with their values, preferences, and individual circumstances.

Feedback and Quality Improvement Processes:

Establish mechanisms for collecting patient feedback, experiences, and satisfaction with shared decision-making processes and patient-centered care.

Use patient feedback to inform quality improvement initiatives, enhance care delivery processes, and address areas for improvement in shared decision-making practices.

By implementing strategies that promote shared decision-making and patient-centered care, healthcare providers and organizations can foster collaborative

139

partnerships with patients, enhance treatment outcomes, improve patient satisfaction, and promote a more patient-centered healthcare system that prioritizes individual needs, preferences, and values. Through collaborative efforts to integrate shared decision-making principles into clinical practice, stakeholders can drive meaningful improvements in access, affordability, and effectiveness in healthcare delivery across the United States.

CHAPTER 8

OVERCOMING POLITICAL AND ECONOMIC CHALLENGES IN HEALTHCARE REFORM

Bipartisan Collaboration and Consensus Building

Foster bipartisan collaboration and consensus building among policymakers, stakeholders, and healthcare experts to develop comprehensive healthcare reform initiatives that address the diverse needs and priorities of all stakeholders.

Encourage dialogue, negotiation, and compromise to overcome political gridlock and partisan divisions, focusing on shared goals of improving access, affordability, and quality of care for all Americans.

Policy Innovation and Experimentation:

Embrace policy innovation and experimentation at the state and federal levels to test alternative models of

141

healthcare delivery, payment reform, and insurance coverage options.

Encourage states to serve as laboratories for innovation by implementing pilot programs, demonstration projects, and waivers to explore new approaches to healthcare financing, delivery, and regulation.

Evidence-Based Policymaking and Data-Driven Solutions:

Promote evidence-based policymaking and data-driven solutions that leverage research, analytics, and evaluation to inform decision-making, assess policy impacts, and identify effective strategies for improving healthcare access and affordability.

Invest in health services research, health economics, and health policy analysis to generate actionable insights and evidence to guide policymakers in designing and implementing reform efforts.

Addressing Cost Drivers and Healthcare Spending:

Address cost drivers and healthcare spending by implementing policies and initiatives aimed at reducing administrative overhead, controlling healthcare prices,

142

and promoting value-based care models that prioritize quality, efficiency, and patient outcomes.

Explore strategies to enhance price transparency, promote competition, and contain costs throughout the healthcare system, including pharmaceutical pricing reforms, payment reforms, and initiatives to eliminate waste and inefficiencies.

Engagement of Stakeholders and Public Participation:

Engage stakeholders, including patients, providers, insurers, employers, advocacy groups, and community organizations, in the healthcare reform process to ensure diverse perspectives are considered and valued.

Promote public participation through town hall meetings, stakeholder forums, and public hearings to solicit input, feedback, and support for healthcare reform initiatives, fostering a sense of ownership and accountability among the public.

Addressing Social Determinants of Health and Health Inequities:

Prioritize efforts to address social determinants of health and health inequities by investing in upstream interventions, community-based initiatives, and cross-sectoral collaborations that address root causes of disparities and promote health equity.

Recognize the interplay between social, economic, and environmental factors in shaping health outcomes, and advocate for policies that address structural barriers, promote social justice, and advance health equity for all populations.

Long-Term Planning and Sustainable Financing:

Engage in long-term planning and sustainable financing strategies to ensure the viability and stability of healthcare reform efforts over time.

Explore innovative financing mechanisms, such as public-private partnerships, value-based payment models, and alternative funding sources, to support healthcare infrastructure investments, workforce development, and population health initiatives.

By overcoming political and economic challenges through collaborative leadership, evidence-based policymaking, stakeholder engagement, and innovative solutions, policymakers and stakeholders can advance meaningful healthcare reform efforts that improve access, affordability, and quality of care for individuals and communities across the United States. Through concerted efforts to address systemic barriers and promote health equity, stakeholders can work towards a more inclusive and sustainable healthcare system that meets the needs of all Americans, now and in the future.

Political Polarization and Its Impact on Healthcare Policy

Partisan Gridlock and Legislative Stalemate:

Political polarization often leads to partisan gridlock and legislative stalemate, hindering efforts to enact comprehensive healthcare reform legislation.

Divisions along party lines can prevent lawmakers from reaching consensus on key healthcare issues, such as insurance coverage expansion, Medicaid expansion, and healthcare financing reforms.

Policy Fragmentation and Incrementalism:

Political polarization may result in policy fragmentation and incrementalism, where piecemeal reforms are pursued instead of comprehensive solutions.

Incremental policymaking may lead to patchwork approaches to healthcare reform, with limited impact on addressing systemic challenges and disparities in access and affordability.

Erosion of Trust in Institutions and Leadership:

Political polarization can contribute to the erosion of trust in government institutions, elected officials, and public leadership, undermining public confidence in the ability of policymakers to address pressing healthcare issues.

Distrust in political institutions may hinder public support for healthcare reform efforts and impede collaboration across partisan lines.

Ideological Battles and Policy Ideologies:

Ideological battles and policy ideologies often shape healthcare policy debates,

146

with competing visions of the role of government, market forces, and individual responsibility in healthcare delivery and financing.

Divergent ideological perspectives on healthcare reform, such as the role of private insurance versus government intervention, can lead to ideological deadlock and ideological purity tests within political parties.

Polarized Media Narratives and Public Discourse:

Polarized media narratives and partisan echo chambers in the media landscape can reinforce ideological divides and shape public discourse on healthcare policy issues.

Biased media coverage and selective reporting may amplify political polarization, deepen ideological divides, and distort public perceptions of healthcare reform proposals and their potential impacts.

Stakeholder Divisions and Interest Group Politics:

Political polarization may exacerbate divisions among stakeholders and interest groups in the healthcare sector, making it challenging to build consensus and mobilize support for reform initiatives.

Special interest politics and lobbying efforts from healthcare industry stakeholders, including insurers, pharmaceutical companies, and provider organizations, can further entrench partisan divisions and influence policymaking outcomes.

State-Level Variation and Policy Experimentation:

Political polarization can contribute to state-level variation in healthcare policy approaches, with Republican-led and Democratic-led states adopting divergent strategies for addressing healthcare access and affordability.

State-level experimentation and innovation in healthcare policy may reflect ideological differences, with states pursuing different paths to reform based on partisan priorities and policy preferences.

Impact on Vulnerable Populations and Health Disparities:

Political polarization and policy gridlock can have disproportionate impacts on vulnerable populations and exacerbate health disparities, particularly for marginalized communities with limited access to healthcare services and resources.

Inadequate or delayed healthcare reform efforts may perpetuate inequities in access to care, exacerbate health outcomes disparities, and widen the gap between the insured and uninsured populations.

Understanding the dynamics of political polarization and its impact on healthcare policy is essential for identifying opportunities for bipartisan cooperation, bridging ideological divides, and advancing meaningful healthcare reform efforts that improve access, affordability, and quality of care for all Americans. By fostering dialogue, collaboration, and compromise across partisan lines, policymakers and stakeholders can work towards building a more inclusive and sustainable healthcare system that meets the needs of diverse populations and promotes health equity.

Economic Implications of Healthcare Reform:

Cost Containment and Fiscal Sustainability:

Healthcare reform initiatives aim to contain rising healthcare costs and promote fiscal sustainability by implementing measures to improve efficiency, reduce waste, and enhance value in healthcare delivery.

Sustainable healthcare financing models, payment reforms, and cost-control mechanisms are essential for addressing budgetary pressures and ensuring long-term affordability of healthcare services.

Impact on Government Spending and Budget Deficits:

Healthcare reform policies have significant implications for government spending and budget deficits, as public expenditures on healthcare programs, such as Medicare and Medicaid, constitute a substantial portion of the federal budget.

Reforms aimed at expanding insurance coverage, improving access to care, and enhancing healthcare quality may require additional government funding or revenue streams to offset costs and mitigate budgetary impacts.

Employer-Sponsored Health Insurance and Labor Market Dynamics:

Changes to employer-sponsored health insurance policies and regulations can influence labor market dynamics, employment decisions, and compensation structures for employers and employees.

Reforms that mandate employer-provided coverage or impose employer mandates may affect business operations, hiring practices, and labor market flexibility, impacting job creation, wages, and employee benefits.

Healthcare Industry Transformation and Market Competition:

Healthcare reform efforts drive industry transformation and reshape market dynamics by promoting competition, innovation, and consumer choice in healthcare markets.

Market-based reforms, such as health insurance exchanges, accountable care organizations, and value-based payment

151

models, incentivize providers to improve efficiency, enhance quality, and respond to consumer preferences.

Investments in Health Information Technology and Infrastructure:

Healthcare reform initiatives often involve investments in health information technology (HIT) infrastructure, electronic health records (EHRs), and interoperable systems to support data exchange, care coordination, and population health management.

Upgrading HIT infrastructure and adopting interoperable EHR platforms require substantial capital investment and ongoing maintenance costs, which can impact healthcare organizations' budgets and financial sustainability.

Healthcare Workforce Development and Training:

Healthcare reform efforts may include initiatives to expand the healthcare workforce, address workforce shortages, and enhance training programs to meet growing demand for healthcare services.

Investments in workforce development, education, and training programs are essential for building a skilled healthcare workforce, addressing workforce shortages in underserved areas, and promoting workforce diversity and cultural competence.

Implications for Health Insurance Markets and Premiums:

Changes to health insurance regulations, coverage requirements, and market rules can influence insurance premiums, cost-sharing arrangements, and consumer affordability in health insurance markets. Reforms aimed at stabilizing insurance markets, expanding risk pools, and reducing adverse selection help mitigate premium volatility, enhance market stability, and improve access to affordable coverage options.

Economic Growth and Productivity:

Access to affordable healthcare services and insurance coverage is critical for promoting economic growth, productivity, and workforce participation by ensuring

153

that individuals and families can access timely and necessary medical care without facing financial hardship.

Healthcare reform efforts that improve health outcomes, reduce healthcare costs, and enhance population health contribute to a healthier, more productive workforce and support overall economic prosperity.

Understanding the economic implications of healthcare reform is essential for policymakers, stakeholders, and the public to evaluate the costs, benefits, trade-offs, and sustainability of proposed reform measures. By considering the economic consequences of healthcare reform initiatives, policymakers can develop evidence-based policies that promote access, affordability, and quality of care while addressing fiscal constraints and economic challenges in the healthcare system.

154

Strategies for Building Consensus and Bipartisan Support in Healthcare Reform

Focus on Shared Goals and Common Ground:

Emphasize shared goals and common ground among policymakers, stakeholders, and political parties, highlighting overarching objectives such as improving access, affordability, and quality of care for all Americans.

Frame healthcare reform discussions in terms of shared values, principles, and aspirations, fostering a sense of unity and collaboration across partisan lines.

Engage Stakeholders and Seek Input:

Engage a diverse range of stakeholders, including patients, providers, insurers, employers, advocacy groups, and community organizations, in the healthcare reform process to solicit input, feedback, and perspectives from all relevant stakeholders.

155

Foster inclusive dialogue, collaboration, and participation to ensure that diverse voices and perspectives are heard and valued in the policymaking process.

Promote Evidence-Based Policymaking:

Promote evidence-based policymaking and data-driven decision-making by providing policymakers with accurate, reliable, and objective information on healthcare issues, trends, and policy options.

Encourage the use of research, analysis, and evaluation to inform policy discussions, assess the potential impacts of reform proposals, and identify effective strategies for achieving shared healthcare goals.

Seek Compromise and Middle Ground Solutions:

Encourage policymakers to seek compromise and middle ground solutions that bridge ideological divides, balance competing interests, and advance incremental progress towards healthcare reform objectives.

Prioritize pragmatic, feasible policy solutions that have the potential to garner bipartisan support and address immediate

156

needs while laying the groundwork for broader reforms over time.

Build Trust and Foster Relationships:

Build trust, establish rapport, and foster constructive relationships among policymakers, stakeholders, and political leaders through regular communication, transparency, and mutual respect.

Cultivate an environment of trust, collaboration, and goodwill by demonstrating a commitment to bipartisanship, integrity, and accountability in the healthcare reform process.

Highlight Success Stories and Best Practices:

Highlight success stories, best practices, and lessons learned from healthcare reform efforts at the state and local levels, showcasing examples of bipartisan cooperation, innovative solutions, and positive outcomes.

Share real-world examples of effective policies, programs, and initiatives that have improved access, affordability, and quality of care in diverse healthcare settings and communities.

157

Empower Grassroots Advocacy and Public Engagement:

Empower grassroots advocacy efforts and public engagement campaigns to mobilize public support for healthcare reform priorities, raise awareness of key issues, and hold policymakers accountable to their constituents.

Foster a sense of civic engagement, activism, and empowerment among citizens, encouraging them to participate in advocacy activities, contact elected officials, and advocate for policies that align with their healthcare needs and priorities.

Focus on Incremental Progress and Long-Term Vision:

Focus on incremental progress and long-term vision in healthcare reform efforts, recognizing that meaningful change often requires sustained effort, persistence, and adaptability over time.

Set realistic short-term goals and benchmarks for progress while maintaining a commitment to broader, transformative reforms that address systemic challenges and promote equity in healthcare delivery.

By employing strategies that prioritize consensus-building, bipartisanship, and collaboration, policymakers and stakeholders can foster a constructive dialogue, advance shared healthcare goals, and drive meaningful improvements in access, affordability, and quality of care for individuals and communities across the United States. Through collective action and inclusive policymaking processes, stakeholders can work together to overcome partisan divisions and build a more equitable and sustainable healthcare system that meets the needs of all Americans.

CHAPTER 9

IMPLEMENTING SUSTAINABLE SOLUTIONS

Long-Term Planning and Vision:

Develop a comprehensive long-term vision for healthcare reform that prioritizes sustainability, equity, and effectiveness in healthcare delivery.

Establish clear goals, objectives, and performance metrics to guide implementation efforts and measure progress over time.

Multi-Stakeholder Collaboration:

Foster multi-stakeholder collaboration and engagement among policymakers, healthcare providers, insurers, employers, patients, advocacy groups, and community organizations.

Create platforms for dialogue, partnership, and shared decision-making to ensure that diverse perspectives are considered and integrated into reform efforts.

Evidence-Based Policy Design:

Base policy design and implementation on rigorous evidence, research, and evaluation to identify effective strategies, interventions, and best practices.

Use data-driven insights to inform decision-making, assess program effectiveness, and adapt interventions based on real-world outcomes and experiences.

Flexibility and Adaptability:

Build flexibility and adaptability into healthcare reform initiatives to accommodate changing healthcare needs, market dynamics, and policy priorities.

Allow for iterative adjustments, course corrections, and refinements based on evolving circumstances, stakeholder feedback, and emerging evidence.

Innovative Financing Mechanisms:

Explore innovative financing mechanisms and sustainable funding sources to support healthcare reform initiatives and investments in healthcare infrastructure, workforce development, and population health.

Consider alternative payment models, public-private partnerships, and value-

161

based purchasing arrangements that align incentives, promote efficiency, and drive value in healthcare delivery.

Health Information Technology (HIT) Integration:

Integrate health information technology (HIT) systems, interoperable electronic health records (EHRs), and data analytics capabilities to support information exchange, care coordination, and population health management.

Invest in HIT infrastructure, data standards, and cybersecurity measures to ensure secure, reliable, and interoperable data sharing across healthcare settings and stakeholders.

Community-Based Approaches:

Prioritize community-based approaches to healthcare reform that engage local stakeholders, address social determinants of health, and promote health equity.

Support community health initiatives, preventive services, and public health interventions that target high-risk populations, reduce disparities, and improve health outcomes at the grassroots level.

Patient-Centered Care Principles:

Embed patient-centered care principles into healthcare delivery models, policies, and practices to empower patients, enhance patient-provider communication, and improve health outcomes.

Prioritize patient engagement, shared decision-making, and care coordination to ensure that healthcare services are responsive to patients' needs, preferences, and priorities.

Continuous Quality Improvement:

Foster a culture of continuous quality improvement (CQI) in healthcare organizations and systems, promoting transparency, accountability, and excellence in care delivery.

Implement quality improvement initiatives, performance metrics, and feedback mechanisms to monitor progress, identify areas for improvement, and drive continuous innovation in healthcare delivery.

Legislative and Regulatory Support:

Provide legislative and regulatory support for sustainable healthcare reform

163

initiatives, including policies that incentivize value-based care, support healthcare innovation, and remove barriers to care access and affordability.

Advocate for policies that strengthen healthcare infrastructure, workforce capacity, and public health preparedness to ensure the resilience and sustainability of the healthcare system in the face of future challenges.

By implementing sustainable solutions that prioritize collaboration, evidence-based policy design, innovation, and patient-centered care, stakeholders can drive meaningful improvements in access, affordability, and quality of care for individuals and communities across the United States. Through collective action and a commitment to long-term sustainability, stakeholders can build a more equitable, resilient, and effective healthcare system that meets the evolving needs of society.

Long-Term Vision for a Reformed Healthcare System

Universal Access to Quality Care:

Our long-term vision for a reformed healthcare system is one where every individual in the United States has universal access to high-quality, affordable healthcare services, regardless of their socioeconomic status, geographic location, or health status.

We envision a healthcare system that ensures equitable access to preventive care, primary care, specialty services, mental health services, and long-term care for all Americans, promoting health and wellness across the lifespan.

Comprehensive Coverage and Benefits:

Our vision includes comprehensive health coverage and benefits that address the full spectrum of healthcare needs, including medical, dental, vision, and prescription drug coverage, as well as mental health and substance abuse treatment services.

165

We advocate for coverage that encompasses preventive services, chronic disease management, maternity care, pediatric care, and end-of-life care, ensuring that individuals receive the care they need to maintain optimal health and well-being.

Patient-Centered Care and Shared Decision-Making:

Central to our vision is a healthcare system that prioritizes patient-centered care, respecting individuals' autonomy, dignity, and preferences in healthcare decision-making.

We envision healthcare providers who partner with patients, families, and caregivers in shared decision-making, empowering individuals to actively participate in their healthcare journey, set treatment goals, and make informed choices about their care.

Value-Based Care and Population Health Management:

Our vision includes a shift towards value-based care models and population health management approaches that prioritize outcomes, efficiency, and cost-effectiveness in healthcare delivery.

166

We advocate for payment reforms that reward quality, promote care coordination, and incentivize providers to focus on preventive care, chronic disease management, and addressing social determinants of health.

Health Equity and Elimination of Disparities:

We are committed to achieving health equity and eliminating disparities in healthcare access, outcomes, and experiences across diverse populations.

Our vision includes targeted interventions and investments that address systemic barriers, structural inequities, and social determinants of health, promoting equity in healthcare access, delivery, and outcomes for all individuals and communities.

Innovation and Technological Advancements:

We embrace innovation and technological advancements as catalysts for transforming healthcare delivery, improving outcomes, and enhancing patient experiences.

Our vision includes harnessing the power of digital health, telemedicine, artificial intelligence, precision medicine, and

167

genomics to personalize care, streamline processes, and revolutionize healthcare delivery models.

Sustainability and Resilience:

We prioritize sustainability and resilience in our healthcare system, ensuring that reforms are financially sustainable, adaptable to changing demographics and healthcare needs, and resilient to economic, environmental, and public health challenges.

Our vision includes investments in healthcare infrastructure, workforce development, and public health preparedness to build a robust and resilient healthcare system that can effectively respond to emergent threats and crises.

Collaboration and Continuous Improvement:

Central to our vision is a culture of collaboration, continuous improvement, and accountability among stakeholders, policymakers, healthcare providers, and communities.

We envision a healthcare system where stakeholders work together to address shared challenges, innovate new solutions, and continuously improve care delivery processes, outcomes, and experiences for individuals and populations.

In pursuing this long-term vision for a reformed healthcare system, we acknowledge that achieving meaningful change will require sustained commitment, bipartisan cooperation, and collective action from all stakeholders. By working together towards common goals, we can build a healthcare system that embodies our values of access, affordability, quality, equity, and compassion, ensuring that every individual in the United States has the opportunity to live a healthy and fulfilling life.

Strategies for Effective Implementation and Evaluation

Clear Goals and Objectives:

Define clear, measurable goals and objectives for healthcare reform initiatives, specifying desired outcomes, target populations, and timelines for implementation.

Ensure alignment between reform efforts and overarching healthcare priorities, such as improving access, affordability, quality, and equity in healthcare delivery.

Stakeholder Engagement and Buy-In:

Engage a broad range of stakeholders, including policymakers, healthcare providers, insurers, patients, advocacy groups, and community organizations, in the implementation process.

Foster collaboration, communication, and transparency to build consensus, generate buy-in, and ensure that diverse perspectives are considered and valued throughout the reform process.

Interdisciplinary Collaboration:

Foster interdisciplinary collaboration and coordination among healthcare professionals, public health experts, social service providers, educators, researchers, and policymakers to address complex healthcare challenges comprehensively.

Facilitate cross-sectoral partnerships and information sharing to leverage expertise, resources, and best practices from diverse disciplines and fields.

Evidence-Based Practice and Continuous Learning:

Base implementation strategies and interventions on evidence-based practices, research findings, and evaluation data to inform decision-making and guide programmatic improvements.

Establish mechanisms for ongoing monitoring, evaluation, and feedback to assess program effectiveness, identify areas for improvement, and adapt strategies in real time.

Capacity Building and Workforce Development:

Invest in workforce development, training, and capacity building to equip healthcare professionals, administrators, and staff with the knowledge, skills, and resources needed to effectively implement healthcare reform initiatives.

Provide opportunities for professional development, leadership training, and interdisciplinary collaboration to enhance workforce readiness and adaptability in a changing healthcare landscape.

171

Resource Allocation and Financial Management:

Allocate resources strategically and prioritize investments in areas that have the greatest potential to impact healthcare access, affordability, and quality.

Implement robust financial management practices, budgetary controls, and performance monitoring mechanisms to ensure accountability, transparency, and fiscal sustainability in healthcare spending.

Technology Integration and Health Information Systems:

Integrate health information technology (HIT) systems, electronic health records (EHRs), and data analytics platforms to support information exchange, care coordination, and population health management.

Invest in interoperable HIT infrastructure, cybersecurity measures, and data governance frameworks to safeguard patient privacy, enhance data integrity, and facilitate data-driven decision-making.

Community Engagement and Empowerment:

Engage communities, patients, and families as partners in the implementation process, soliciting input, feedback, and participation in healthcare reform initiatives.

Empower individuals and communities to advocate for their healthcare needs, access resources, and participate in decision-making processes that affect their health and well-being.

Policy Advocacy and Public Education:

Advocate for supportive policies, legislation, and regulatory reforms that facilitate effective implementation of healthcare reform initiatives and address systemic barriers to access and affordability.

Conduct public education campaigns, outreach activities, and community forums to raise awareness, promote health literacy, and empower individuals to navigate the healthcare system effectively.

Continuous Quality Improvement and Innovation:

Embrace a culture of continuous quality improvement (CQI) and innovation in healthcare delivery, encouraging experimentation, learning, and adaptation to drive positive change.

Foster a supportive environment for innovation, risk-taking, and creative problem-solving, recognizing and rewarding individuals and teams that develop innovative solutions to healthcare challenges.

By implementing these strategies for effective implementation and evaluation, stakeholders can maximize the impact of healthcare reform initiatives, achieve meaningful improvements in access and affordability, and advance the shared goal of building a more equitable, efficient, and sustainable healthcare system for all Americans.

174

Ensuring Sustainability and Adaptability in the Face of Future Challenges

Anticipating Demographic Shifts and Changing Healthcare Needs:

Proactively anticipate demographic shifts, population aging, and changing healthcare needs to adapt healthcare delivery models, services, and infrastructure accordingly.

Consider the impact of population growth, migration patterns, and evolving disease trends on healthcare demand and resource allocation.

Building Resilience to Public Health Crises and Emergencies:

Strengthen healthcare system resilience to mitigate the impact of public health crises, pandemics, natural disasters, and other emergencies.

Invest in disaster preparedness, surge capacity, and emergency response capabilities to ensure timely and effective coordination of healthcare services during crises.

175

Addressing Technological Advances and Health Innovations:

Embrace technological advances, health innovations, and digital transformation in healthcare delivery, leveraging telemedicine, artificial intelligence, remote monitoring, and other emerging technologies.

Stay abreast of evolving best practices, evidence-based interventions, and transformative trends in healthcare innovation to enhance patient care, improve outcomes, and optimize resource utilization.

Navigating Regulatory and Policy Changes:

Navigate regulatory and policy changes at the local, state, and federal levels by staying informed, engaging in advocacy efforts, and actively participating in policy dialogue.

Monitor legislative developments, policy debates, and regulatory updates to anticipate potential impacts on healthcare operations, financing, and reimbursement.

Promoting Fiscal Sustainability and Resource Stewardship:

Promote fiscal sustainability and resource stewardship in healthcare spending by adopting cost-effective practices, prioritizing investments in high-value care, and eliminating waste and inefficiencies.

Implement budgetary controls, performance metrics, and outcome measures to track spending, monitor financial performance, and optimize resource allocation across healthcare systems and organizations.

Cultivating a Culture of Learning and Adaptation:

Cultivate a culture of learning, adaptation, and continuous improvement within healthcare organizations, fostering a spirit of innovation, collaboration, and resilience among staff and leadership.

Encourage staff engagement, professional development, and knowledge sharing to foster a culture of continuous learning and adaptability in response to evolving healthcare challenges.

Engaging Stakeholders and Communities:

Engage stakeholders, patients, families, and communities as partners in healthcare reform efforts, soliciting input, feedback, and participation in decision-making processes.

Foster transparent communication, trust-building, and collaborative problem-solving to address community needs, preferences, and priorities in healthcare service delivery.

Embracing Flexibility and Agility in Policy and Practice:

Embrace flexibility and agility in policy formulation, implementation, and adaptation to accommodate changing healthcare landscapes, market dynamics, and regulatory environments.

Advocate for policies and regulations that support innovation, experimentation, and creative problem-solving in healthcare delivery, payment models, and quality improvement initiatives.

By ensuring sustainability and adaptability in the face of future challenges, stakeholders can build a resilient, responsive, and patient-centered

178

healthcare system that meets the evolving needs of individuals and communities across the United States. Through proactive planning, collaboration, and innovation, stakeholders can navigate uncertainty, seize opportunities, and drive positive change in healthcare delivery, access, and affordability for generations to come.

CHAPTER 10

CONCLUSION

In "Healthcare Reform: Solutions for Improving Access and Affordability in the US," we have explored the multifaceted challenges facing our healthcare system and presented comprehensive strategies for meaningful reform. Throughout our journey, we have underscored the urgent need to address issues of access, affordability, quality, and equity in healthcare delivery to ensure the well-being of all Americans.

The United States stands at a critical juncture, where the imperatives of access and affordability intersect with the demands of an ever-evolving healthcare landscape. Our current system is characterized by disparities in care, rising costs, and systemic inefficiencies that undermine the health and financial security of millions of individuals and families. Yet, within these challenges lie opportunities for transformative change and innovation.

We have examined the historical context of healthcare reform, analyzed the complexities of our healthcare system, and identified the root causes of access and affordability issues. From understanding the social determinants of health to evaluating the impact of insurance coverage gaps, we have delved into the intricacies of healthcare delivery and financing.

Importantly, we have not only highlighted problems but also presented actionable solutions. From policy approaches to innovation in healthcare delivery, we have explored a spectrum of strategies aimed at improving access, enhancing affordability, and promoting patient-centered care. We have championed collaborative efforts, evidence-based policymaking, and stakeholder engagement as essential pillars of successful reform.

As we conclude our exploration, it is clear that the path to meaningful healthcare reform is neither simple nor linear. It requires political will, collective action, and sustained commitment from all stakeholders. It demands a willingness to confront entrenched interests, challenge the status quo, and prioritize the well-being of patients above all else.

Moving forward, we must remain steadfast in our pursuit of a healthcare system that

181

reflects our values of equity, compassion, and justice. We must continue to advocate for policies that expand access, reduce disparities, and promote affordability for all Americans, regardless of race, ethnicity, income, or geography.

In the face of uncertainty and complexity, let us draw inspiration from the resilience and innovation of healthcare providers, advocates, and communities across the nation. Let us harness the power of collaboration, innovation, and compassion to build a healthcare system that works for everyone one that honors the dignity and humanity of every individual we serve.

Together, we can chart a course towards a brighter, healthier future one where access to quality care is not a privilege but a fundamental right. As we embark on this journey of transformation, let us remain steadfast in our commitment to healthcare reform and unwavering in our pursuit of a healthier, more equitable America.

The time for change is now. Let us seize this moment and forge a path towards a healthcare system that reflects our highest ideals and aspirations. The future of healthcare reform lies in our hands, and together, we have the power to shape a better tomorrow for generations to come.

Recap of Key Insights and Recommendations

Throughout "Healthcare Reform: Solutions for Improving Access and Affordability in the US," we have explored the complex landscape of healthcare delivery, access, and affordability. Drawing on extensive research, analysis, and stakeholder perspectives, we have identified key insights and recommendations aimed at addressing the pressing challenges facing our healthcare system. Here is a recap of our most salient insights and recommendations:

Understanding the Current Landscape: We began by examining the historical context of healthcare reform in the US and the factors contributing to access and affordability issues. From insurance coverage gaps to barriers faced by different demographics, we gained a comprehensive understanding of the systemic challenges at play.

Barriers to Access and Affordability:
We identified various barriers to access, including geographical disparities, financial constraints, and social determinants of health. From rural populations to low-income individuals, we recognized the diverse needs and challenges faced by different segments of the population.

Impact of Insurance Coverage Gaps:
We explored the implications of insurance coverage gaps on healthcare access and affordability, highlighting the importance of expanding coverage and reducing disparities in insurance enrollment and access to care.

Affordability in Healthcare: We delved into rising healthcare costs and their implications for patients and families, emphasizing the need for innovative solutions to curb costs while ensuring quality care remains accessible and affordable.

Policy Approaches to Reform:
We discussed various policy approaches to healthcare reform, including expanding Medicaid, strengthening the Affordable

184

Care Act (ACA), and exploring alternative payment models. We emphasized the importance of bipartisan collaboration and evidence-based policymaking in driving meaningful reform.

Innovation in Healthcare Delivery:

We explored the role of innovation in healthcare delivery, including telemedicine, value-based care models, and preventive healthcare initiatives. We underscored the potential of technology and data-driven approaches to improve access, enhance quality, and reduce costs.

Addressing Social Determinants of Health:

We highlighted the impact of social factors on healthcare outcomes and advocated for community-based interventions, collaborative efforts between healthcare providers and social service organizations, and patient education and advocacy initiatives to address social determinants of health.

Ensuring Sustainability and Adaptability:

We emphasized the importance of ensuring sustainability and adaptability in the face of future challenges, including demographic shifts,

185

technological advancements, and regulatory changes. We underscored the need for proactive planning, stakeholder engagement, and continuous learning to navigate uncertainty and drive positive change.

In conclusion, "Healthcare Reform: Solutions for Improving Access and Affordability in the US" offers a comprehensive framework for advancing healthcare reform efforts. By prioritizing access, affordability, quality, and equity in healthcare delivery, and by embracing innovation, collaboration, and evidence-based policymaking, we can work towards building a more accessible, affordable, and equitable healthcare system for all Americans. The insights and recommendations presented in this book provide a roadmap for policymakers, healthcare providers, advocates, and communities to drive meaningful change and improve the health and well-being of individuals and families across the nation.

Call to Action

In "Healthcare Reform: Solutions for Improving Access and Affordability in the US," we have outlined the challenges and

186

opportunities facing our healthcare system and presented actionable strategies for reform. Now, it is time for stakeholders across the nation to prioritize healthcare reforms and drive meaningful change. Here is our call to action:

Commit to Bipartisan Collaboration:

We call on policymakers, elected officials, and political leaders to set aside partisan differences and prioritize bipartisan collaboration in advancing healthcare reform efforts. The health and well-being of our nation's citizens should transcend political ideologies, and meaningful progress requires cooperation and compromise across party lines.

Listen to Stakeholder Voices: We urge healthcare providers, insurers, patients, advocacy groups, and community organizations to actively engage in the reform process and make their voices heard. Your insights, experiences, and perspectives are invaluable in shaping policies and practices that reflect the needs and priorities of diverse communities.

Advocate for Equitable Access and Affordability: We encourage advocates and activists to raise awareness about the importance of equitable access and affordability in healthcare. Use your platforms to amplify the voices of marginalized and underserved populations, advocate for policy reforms that address systemic disparities, and hold decision-makers accountable to their commitments to health equity.

Embrace Innovation and Evidence-Based Practices: We challenge healthcare professionals, researchers, and innovators to embrace cutting-edge technologies, evidence-based practices, and transformative approaches to healthcare delivery. Innovations in telemedicine, digital health, precision medicine, and population health management have the potential to revolutionize care delivery and improve outcomes for patients across the continuum of care.

Invest in Community-Based Solutions: We call on local governments, philanthropic organizations, and community leaders to invest in community-based solutions that address social determinants of health, promote health equity, and empower individuals to take control of their health and well-being. By investing in upstream interventions and preventive care initiatives, we can reduce disparities and improve health outcomes for all.

Hold Stakeholders Accountable:

We urge all stakeholders to hold themselves and others accountable to the principles of transparency, equity, and patient-centered care. Demand accountability from healthcare providers, insurers, policymakers, and industry stakeholders to ensure that reform efforts are implemented effectively and equitably.

Sustain Momentum for Change: Finally, we encourage all stakeholders to sustain the momentum for change by remaining engaged, informed, and committed to the long-term vision of a healthcare system that works for everyone. Reforming our healthcare system is a marathon, not a sprint, and it requires sustained effort, collaboration, and advocacy from all of us.

Together, we have the power to transform our healthcare system and create a future where access to high-quality, affordable care is a fundamental right for all Americans. Let us seize this opportunity to

189

prioritize healthcare reforms, drive positive change, and build a healthier, more equitable future for generations to come. The time for action is now.

Vision for a More Accessible and Affordable Healthcare System in the US

Our vision for a more accessible and affordable healthcare system in the United States is one that prioritizes the health and well-being of every individual, regardless of their socioeconomic status, geographic location, or health condition. This vision encompasses the following key principles and objectives:

Universal Access to Care: Our healthcare system must ensure universal access to comprehensive healthcare services for all Americans, irrespective of their ability to pay or their pre-existing health conditions. Every individual should have the opportunity to receive timely, high-quality care when they need it, without facing financial barriers or disparities in access.

Equitable Distribution of Resources: We envision a healthcare system that allocates resources equitably, prioritizing underserved communities, rural populations, and marginalized groups that have historically faced barriers to access. Investments in healthcare infrastructure, workforce development, and preventive care initiatives should be targeted towards addressing disparities and promoting health equity.

Affordable Coverage Options: Affordable coverage options must be made available to individuals and families, with subsidies and financial assistance provided to those who need it most. Insurance plans should offer comprehensive benefits at reasonable premiums, deductibles, and copayments, ensuring that healthcare remains affordable and accessible to all.

Value-Based Care and Prevention: Our healthcare system should transition towards value-based care models that prioritize prevention, early intervention, and chronic disease management. By incentivizing healthcare providers to focus on preventive services,

191

wellness programs, and population health management, we can improve health outcomes while reducing long-term healthcare costs.

Innovative Delivery Models:

Embracing innovative delivery models such as telemedicine, mobile clinics, and community health centers can expand access to care, especially in underserved areas and remote regions. These models leverage technology and community partnerships to reach patients where they are, overcoming geographical barriers and increasing healthcare access.

Transparency and Accountability:

Transparency and accountability should be central tenets of our healthcare system, with clear pricing information, quality metrics, and patient outcomes publicly available to inform consumer choice and drive improvements in care delivery. Healthcare providers and insurers should be held accountable for delivering high-quality, cost-effective care that meets the needs of patients.

Cross-Sector Collaboration:

Achieving our vision requires collaboration across multiple sectors, including healthcare, public health, education, housing, and social services. By addressing the social determinants of health and fostering cross-sector partnerships, we can create healthier communities and reduce the burden of preventable diseases and health disparities.

Continuous Improvement and Adaptation: Our healthcare system must be dynamic, adaptive, and responsive to changing demographics, technological advancements, and emerging health threats. Continuous improvement efforts, informed by data and feedback from patients and providers, should drive innovation and ensure that healthcare delivery remains patient-centered, efficient, and effective.

In realizing this vision for a more accessible and affordable healthcare system, we recognize the challenges ahead and the need for sustained commitment from policymakers, healthcare providers, insurers, advocacy groups, and communities. By working together with a shared sense of purpose and urgency, we can transform our healthcare system into one that embodies the principles of equity,

193

compassion, and excellence, ensuring that every American has the opportunity to lead a healthy and fulfilling life.